Organic Contaminants, Trace and Major Elements, and Nutrients in Water and Sediment Sampled in Response to the Deepwater Horizon Oil Spill

By Lisa H. Nowell, Amy S. Ludtke, David K. Mueller, and Jonathon C. Scott

Scientific Investigations Report 2012–5228

U.S. Department of the Interior
U.S. Geological Survey

U.S. Department of the Interior
KEN SALAZAR, Secretary

U.S. Geological Survey
Suzette M. Kimball, Acting Director

U.S. Geological Survey, Reston, Virginia: 2013

For more information on the USGS—the Federal source for science about the Earth, its natural and living resources, natural hazards, and the environment, visit http://www.usgs.gov or call 1–888–ASK–USGS.

For an overview of USGS information products, including maps, imagery, and publications, visit http://www.usgs.gov/pubprod

To order this and other USGS information products, visit http://store.usgs.gov

Suggested citation:
Nowell, L.H., Ludtke, A.S., Mueller, D.K., and Scott, J.C., 2013, Organic contaminants, trace and major elements, and nutrients in water and sediment sampled in response to the Deepwater Horizon oil spill: U.S. Geological Survey Open-File Report 2012–5228, 96 p, plus appendixes.

Contents

Contents

Figures

Tables

Tables

Tables

Conversion Factors

SI to Inch/Pound

Multiply	By	To obtain
Length		
centimeter (cm)	0.3937	inch (in.)
millimeter (mm)	0.03937	inch (in.)
meter (m)	3.281	foot (ft)
kilometer (km)	0.6214	mile (mi)
kilometer (km)	0.5400	mile, nautical (nmi)
meter (m)	1.094	yard (yd)
Area		
square meter (m^2)	0.0002471	acre
square kilometer (km^2)	0.3861	square mile (mi^2)
Volume		
cubic meter (m^3)	6.290	barrel (petroleum, 1 barrel = 42 gal)
liter (L)	33.82	ounce, fluid (fl. oz)
liter (L)	2.113	pint (pt)
liter (L)	1.057	quart (qt)
liter (L)	0.2642	gallon (gal)
liter (L)	61.02	cubic inch (in^3)
Mass		
gram (g)	0.03527	ounce, avoirdupois (oz)
kilogram (kg)	2.205	pound avoirdupois (lb)

Temperature in degrees Celsius (°C) may be converted to degrees Fahrenheit (°F) as follows:

$$°F=(1.8×°C)+32$$

Temperature in degrees Fahrenheit (°F) may be converted to degrees Celsius (°C) as follows:

$$°C=(°F-32)/1.8$$

Concentrations of chemical constituents in water are given either in milligrams per liter (mg/L), micrograms per liter (µg/L), or percent.

Abbreviations and Acronyms

AET	Apparent effect threshold
BTEX	Benzene, toluene, ethylbenzene, xylene and related volatile (compounds)
DOC	dissolved organic carbon
ERL	Effects Range-Low
ERM	Effects Range-Median
$ESBTU_i$	equilibrium-partitioning sediment benchmark toxic-unit concentration
EqP	equilibrium-partitioning
GC/MS	gas chromatography with mass spectrometry
GOM	Gulf of Mexico
GPS	Global Positioning System
ICP-OES	inductively-coupled plasma—optical emission spectrometry
ICP-MS	inductively-coupled plasma—mass spectrometry
K_{oc}	equilibrium partition coefficient
M-1	Macondo-1
MC252	Mississippi Canyon 252
MDL	method detection limits
PAH	polycyclic aromatic hydrocarbons
PEC	probable effect concentration
PEL	probable effect level
PPW	paired Prentice-Wilcoxon
QC	quality control
RSD	relative standard deviation
SIM	selective ion monitoring mode
SVOC	semivolatile organic compounds
TEC	threshold effect concentration
TEL	threshold effect level
TOC	total organic carbon
TU	toxic unit
VOC	volatile organic compounds

Organizations

API	American Petroleum Institute
BLM	Bureau of Land Management
BP	British Petroleum
NOAA	National Oceanic and Atmospheric Administration
NWQL	USGS National Water Quality Laboratory
OCRL	Organic Carbon Research Laboratory
SCL	USGS Sediment Chemistry Laboratory
USEPA	U.S. Environmental Protection Agency
USGS	U.S. Geological Survey

Units of measurement

mg/kg	milligram per kilogram
µg/g	microgram per gram
µg/kg	microgram per kilogram
µg/L	microgram per liter
µm	micrometer

Organic Contaminants, Trace and Major Elements, and Nutrients in Water and Sediment Sampled in Response to the Deepwater Horizon Oil Spill

By Lisa H. Nowell, Amy S. Ludtke, David K. Mueller, and Jonathon C. Scott

Abstract

Beach water and sediment samples were collected along the Gulf of Mexico coast to assess differences in contaminant concentrations before and after landfall of Macondo-1 well oil released into the Gulf of Mexico from the sinking of the British Petroleum Corporation's Deepwater Horizon drilling platform. Samples were collected at 70 coastal sites between May 7 and July 7, 2010, to document baseline, or "pre-landfall" conditions. A subset of 48 sites was resampled during October 4 to 14, 2010, after oil had made landfall on the Gulf of Mexico coast, called the "post-landfall" sampling period, to determine if actionable concentrations of oil were present along shorelines.

Few organic contaminants were detected in water; their detection frequencies generally were low and similar in pre-landfall and post-landfall samples. Only one organic contaminant—toluene—had significantly higher concentrations in post-landfall than pre-landfall water samples. No water samples exceeded any human-health benchmarks, and only one post-landfall water sample exceeded an aquatic-life benchmark—the toxic-unit benchmark for polycyclic aromatic hydrocarbons (PAH) mixtures. In sediment, concentrations of 3 parent PAHs and 17 alkylated PAH groups were significantly higher in post-landfall samples than pre-landfall samples. One pre-landfall sample from Texas exceeded the sediment toxic-unit benchmark for PAH mixtures; this site was not sampled during the post-landfall period. Empirical upper screening-value benchmarks for PAHs in sediment were exceeded at 37 percent of post-landfall samples and 22 percent of pre-landfall samples, but there was no significant difference in the proportion of samples exceeding benchmarks between paired pre-landfall and post-landfall samples. Seven sites had the largest concentration differences between post-landfall and pre-landfall samples for 15 alkylated PAHs. Five of these seven sites, located in Louisiana, Mississippi, and Alabama, had diagnostic geochemical evidence of Macondo-1 oil in post-landfall sediments and tarballs.

For trace and major elements in water, analytical reporting levels for several elements were high and variable. No human-health benchmarks were exceeded, although these were available for only two elements. Aquatic-life benchmarks for trace elements were exceeded in 47 percent of water samples overall. The elements responsible for the most exceedances in post-landfall samples were boron, copper, and manganese. Benchmark exceedances in water could be substantially underestimated because some samples had reporting levels higher than the applicable benchmarks (such as cobalt, copper, lead and zinc) and some elements (such as boron and vanadium) were analyzed in samples from only one sampling period. For trace elements in whole sediment, empirical upper screening-value benchmarks were exceeded in 57 percent of post-landfall samples and 40 percent of pre-landfall samples, but there was no significant difference in the proportion of samples exceeding benchmarks between paired pre-landfall and post-landfall samples. Benchmark exceedance frequencies could be conservatively high because they are based on measurements of total trace-element concentrations in sediment. In the less than 63-micrometer sediment fraction, one or more trace or major elements were anthropogenically enriched relative to national baseline values for U.S. streams for all sediment samples except one. Sixteen percent of sediment samples exceeded upper screening-value benchmarks for, and were enriched in, one or more of the following elements: barium, vanadium, aluminum, manganese, arsenic, chromium, and cobalt. These samples were evenly divided between the sampling periods.

Aquatic-life benchmarks were frequently exceeded along the Gulf of Mexico coast by trace elements in both water and sediment and by PAHs in sediment. For the most part, however, significant differences between pre-landfall and post-landfall samples were limited to concentrations of PAHs in sediment. At five sites along the coast, the higher post-landfall concentrations of PAHs were associated with diagnostic geochemical evidence of Deepwater Horizon Macondo-1 oil.

Introduction

On April 20, 2010, the British Petroleum (BP) Corporation's Deepwater Horizon Mississippi Canyon 252 (MC252) drilling platform sank following an explosion, and oil and gas began to be released into the northern Gulf of Mexico (GOM) from the ruptured Macondo-1 (M-1) well approximately 5,000 feet below the sea surface. About 4.93 million barrels (205.8 million gallons) of oil were released into the northern GOM by the time the well was successfully capped on July 15, 2010 (Operational Science Advisory Team, 2010). To disperse the oil, 1.84 million gallons of chemical dispersants were applied to surface oil and at the well-head (Operational Science Advisory Team, 2010). In response to the threat of oil affecting sensitive habitat along the shores of the GOM, the U.S. Geological Survey (USGS) collected near-surface beach water and sediment at coastal sites from Texas to Florida, both before and after the oil made landfall on the GOM coast. "Pre-landfall" samples were collected from May 7 to July 7, 2010, and "post-landfall" samples were collected on August 23 and from October 4 to 14, 2010. The post-landfall study was requested by the U.S. Coast Guard (Wilde and Skrobialowski, 2011) and was used in conjunction with data from other sources, including the U.S. Environmental Protection Agency, National Oceanic and Atmospheric Administration, the GOM coast states, and BP, to assess the distribution of actionable—that is, amenable to removal actions—oil-related chemicals that remain in the water column, sediments, or both, and to inform decision makers on further oil-removal operations (Operational Science Advisory Team, 2010; Unified Area Command, 2010).

The purpose of this report is to characterize the water and sediment chemistry in pre-landfall and post-landfall samples and to ascertain whether there were significant changes between the two sampling periods. This report complements activities of other USGS scientists and USGS production and research laboratories who are determining surfactants in water samples; analyzing geochemical markers for the presence of M-1 oil, also called MC252 oil, in sediment and tarballs; evaluating bacterial populations capable of degrading oils; assessing the toxicity of sediment pore water to the sea urchin (*Arbacia punctulata*); and assessing benthic macroinvertebrate indicators of shoreline habitat conditions (Donna N. Myers, Chief, Office of Water Quality, U.S. Geological Survey, Reston, Va., written commun., September 9, 2011).

Specific objectives of this report are the following:

- Summarize the occurrence of organic contaminants, trace and major elements, and nutrients in water and sediment samples at sites along the GOM coast sampled by the USGS before and after oil made landfall.

- Compare contaminant concentrations in pre-landfall samples to post-landfall samples for water and sediment.

- Compare measured concentrations of contaminants to applicable existing benchmarks for protection of human health, aquatic life, and sediment quality.

This report presents one of multiple lines of evidence documenting conditions along the GOM coast before and after landfall of M-1 oil.

Methods

This study had two sample collection periods: pre-landfall and post-landfall. Pre-landfall samples were collected from May 7 to July 7, 2010, which was after the oil spill began, but before oil made landfall on the GOM coast. Post-landfall samples were collected on August 23 and from October 4 to 14, 2010, after oil made landfall at the sampled sites. Post-landfall sampling was carried out at a subset of the pre-landfall sampling sites, plus one oil-affected site that was not sampled during the pre-landfall period.

Although the sample-collection methods were mostly the same during pre-landfall and post-landfall periods, the priorities for chemical analyses changed in some ways between the pre-landfall and post-landfall periods as more information became available from the National Oceanic and Atmospheric Administration (NOAA) and the U.S. Environmental Protection Agency (USEPA) about methods and priorities for oil-related chemical contaminant testing (Operational Science Advisory Team, 2010, appendix F). As a result, the choice of some chemical analysis methods and laboratories was revised for the post-landfall period period. Differences in methods between the two sampling periods are described in the "Chemical Analyses" section.

Study Area and Site Selection

The initial response of the USGS to the Deepwater Horizon oil spill required rapid mobilization to collect water and sediment samples before landfall of the oil in order to establish a baseline chemical and biological profile. This baseline profile could then be used to understand any post-landfall effects on, or changes to, GOM coastal environments (Wilde and Skrobialowski, 2011). The USGS Water Science Centers in Texas, Louisiana, Mississippi, Alabama, and Florida coordinated efforts to sample water and sediments at 70 sites from beaches, barrier islands, and coastal wetlands that could be adversely affected by oil from the spill coming ashore (fig. 1). High priority was given to coastal wetlands, Department of Interior lands at risk for oil contamination, such as National Wildlife Refuges, Bureau of Land Management (BLM) lands, National Seashore areas, and State Parks (Rosenbauer and others, 2010; Donna Myers, Chief, Office of Water Quality, U.S. Geological Survey, Reston, Va., written commun., September 9, 2011). The purpose of the sampling was to define pre-landfall conditions in the physical, chemical, biological, and microbiological quality of the nearshore environment. Pre-landfall samples were collected between May 7 and July 7, 2010 (fig. 1; table 1). Global Positioning System (GPS) coordinates were recorded at each site so that any subsequent samples could be collected in the same location.

Post-landfall sampling was carried out following a request by the U.S. Coast Guard to assess whether actionable levels of Deepwater Horizon-related oil-spill contamination were present after the extensive clean-up efforts of coastal areas by BP (Wilde and Skrobialowski, 2011). Sampling was performed by the same USGS Water Science Centers that collected data for the pre-landfall assessment. Post-landfall samples were collected at 48 of the original 70 pre-landfall sites plus 1 oil-affected wetland site at Bay Jimmy, Louisiana, which was not sampled before landfall, making a total of 71 sites (fig. 1; table 1). Post-landfall sites were selected from among the pre-landfall site locations on the basis of the extent of oil observed at the surface, as ascertained from ships, aircraft, satellites, and *in situ* sampling; knowledge of the nearshore physical oceanography, that is, movement of water and sediments; and trajectory modeling by NOAA (National Oceanic and Atmospheric Adminstration, 2010; Unified Area Command, 2010). The purpose of the post-landfall sampling was to document residual, or actionable, oil.

Sample Collection

One water sample and one composite sediment sample were collected at each pre-landfall and post-landfall site by personnel from the USGS Water Science Centers from the GOM coastal states. All pre-landfall samples were collected between May 7 and July 7, 2010. All post-landfall samples were collected between October 4 and 14, 2010, except the Bay Jimmy site, which was sampled on August 23, 2010 (table 1). Post-landfall sampling took place after the arrival of M-1 oil at the sampled sites. USGS field teams collected pre-landfall and post-landfall samples and site data by following protocols and procedures described in Wilde and Skrobialowski (2011) and in the USGS National Field Manual for the Collection of Water-Quality Data (variously dated). Post-landfall samples at each site were collected at, or close to, the pre-landfall sampling location (Wilde and Skrobialowski, 2011). All samples of water and sediment were collected near the land/water interface, as described in the following sections. Samples were intended to be representative of ambient conditions at the time of sample collection (Wilde and Skrobialowski, 2011). Water samples were collected first and packed in coolers, and then sediment samples were collected. The collected samples were held on ice at 4 degrees Celsius (°C) after collection and during transport under chain-of-custody to the respective laboratories for chemical analysis.

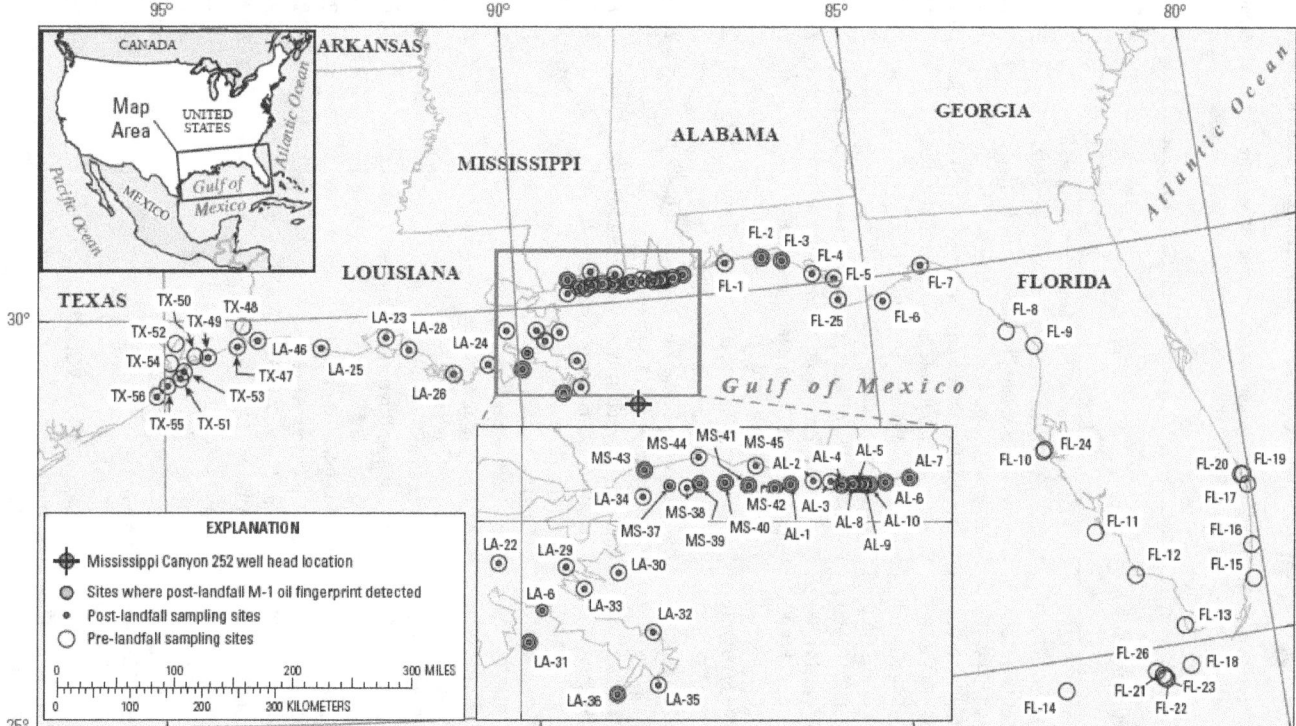

Figure 1. U.S. Geological Survey sites sampled in response to the Deepwater Horizon oil spill, Gulf of Mexico, 2010. Modified from Wilde and Skrobialowski (2011) by Gregory Wetherbee and David Strong

Table 1. U.S. Geological Survey pre-landfall and post-landfall sampling sites in the Gulf of Mexico, 2010 Deepwater Horizon oil spill.

[Sample dates are expressed as MM/DD/YY. **Abbreviations**: Ala, Alabama; BLM, Bureau of Land Management; dd, decimal degrees; Fla., Florida; La., Louisiana; M-1, Macondo-1 well; Miss, Mississippi; NWISweb, online National Water Information System (http://nwis.waterdata.usgs.gov/usa/nwis/qwdata); NWR, National Wildlife Refuge; St., Saint; Tex., Texas; USGS, U.S. Geological Survey; WSC, Water Science Center; –, no post-landfall sample was collected]

Map number	Station identifier	Site name	Latitude (dd)	Longitude (dd)	USGS WSC	Dates that site was sampled Post-landfall	Dates that site was sampled Pre-landfall[2]	M-1 oil in post-landfall[1] Sediment	M-1 oil in post-landfall[1] Tarball
FL-1	302144086581200	Gulf Island National Seashore near Navarre, Fla.	30.36239	-86.9702	Fla.	10/04/10	05/11/10	Mix	No
FL-2	302252086263400	Henderson Beach State Park near Destin, Fla.	30.38294	-86.4428	Fla.	10/05/10	05/11/10	No	Yes
FL-3	301926086091800	Grayton Beach State Park near Seaside, Fla.	30.32406	-86.1551	Fla.	10/05/10	05/12/10	Yes	–
FL-4	300729085440900	St. Andrews State Park near Panama City, Fla.	30.12472	-85.736	Fla.	10/11/10	05/12/10	No	–
FL-5	294645085243000	St. Joseph Peninsula State Park near Port St. Joe, Fla.	29.77917	-85.4085	Fla.	10/13/10	05/13/10	No	–
FL-6	294152084460300	St. George Island State Park near E Point, Fla.	29.69786	-84.7678	Fla.	10/06/10	05/13/10	No	–
FL-7	300427084105000	St. Marks NWR near St. Marks, Fla.	30.07419	-84.1804	Fla.	10/07/10	05/18/10	No	–
FL-8	290740083031200	Piney Point Beach at Cedar Key, Fla.	29.12775	-83.0534	Fla.	–	05/18/10	–	–
FL-9	285425082412600	Fort Island Gulf Beach near Chassahowitzka, Fla.	28.90719	-82.6908	Fla.	–	05/19/10	–	–
FL-10	273728082441800	Fort DeSoto Park near St. Petersburg, Fla.	27.62444	-82.7383	Fla.	–	05/17/10	–	–
FL-11	263132082114000	Captiva Island Beach near Captiva, Fla.	26.52564	-82.1942	Fla.	–	05/20/10	–	–
FL-12	255609081440700	Tiger Tail Beach at Marco Island, Fla.	25.93614	-81.7346	Fla.	–	05/21/10	–	–
FL-13	251329081101100	Northwest Cape Sable Beach near Flamingo, Fla.	25.22481	-81.17	Fla.	–	05/22/10	–	–
FL-14	243737082522500	Dry Tortugas National Park, Fla.	24.62714	-82.8736	Fla.	–	05/20/10	–	–
FL-15	254002080092000	Bill Baggs Cape near Key Biscayne, Fla.	25.66742	-80.1555	Fla.	–	06/01/10	–	–
FL-16	260454080063400	Lloyd Beach at Fort Lauderdale, Fla.	26.08169	-80.1094	Fla.	–	05/26/10	–	–
FL-17	264921080021700	MacArthur Beach at West Palm Beach, Fla.	26.82258	-80.0381	Fla.	–	05/27/10	–	–
FL-18[2]	244345081000600	Coco Plum Beach near Marathon, Fla.	24.72925	-81.17	Fla.	–	05/24/10	–	–
FL-19	265722080045400	BLM Tract1 near Jupiter Inlet, Fla.	26.95611	-80.0817	Fla.	–	06/16/10	–	–
FL-20	265722080045500	BLM Tract2 near Jupiter Inlet, Fla.	26.95611	-80.0819	Fla.	–	06/16/10	–	–
FL-21	243902081332700	BLM Tract1 near Park Key, Fla.	24.65056	-81.5575	Fla.	–	06/09/10	–	–
FL-22	243703081323700	BLM Tract2 near Sugarloaf Key, Fla.	24.6175	-81.5436	Fla.	–	06/09/10	–	–
FL-23	243700081322300	BLM Tract3 near Sugarloaf Key, Fla.	24.61667	-81.5397	Fla.	–	06/09/10	–	–
FL-24	273605082454900	BLM Tract at Egmont Key, Fla.	27.60139	-82.7636	Fla.	–	06/14/10	–	–
FL-25	300223085260800	BLM Lathrop Bayou near Panama City, Fla.	30.03894	-85.4355	Fla.	10/12/10	06/10/10	No	–
FL-26	244325081351500	Marvin Key at Great White Heron NWR, Fla.	24.70981	-81.6446	Fla.	–	07/07/10	–	–
AL-1	301338088193500	West Dauphin Island, Ala.	30.22743	-88.3264	Ala.	10/13/10	05/08/10	Mix	Yes
AL-2	301455088110300	Dauphin Island, AL-2	30.24881	-88.1842	Ala.	10/07/10	05/09/10	No	–
AL-3	301448088044000	Dauphin Island, AL-3	30.24687	-88.0778	Ala.	10/06/10	05/09/10	No	–
AL-4	301329088003000	Fort Morgan, AL-4	30.22493	-88.0083	Ala.	10/12/10	05/08/10	Yes	Yes
AL-5	301349087541600	Fort Morgan, AL-5	30.23048	-87.9044	Ala.	10/13/10	05/08/10	Yes	Yes
AL-6	301428087434900	Gulf Shores, Ala.	30.24131	-87.7303	Ala.	10/14/10	05/08/10	Mix	Yes
AL-7	301608087345400	Orange Beach, Ala.	30.26909	-87.5816	Ala.	10/14/10	05/08/10	Mix	Yes
AL-8	301353087561600	BLM-1, Ala.	30.23159	-87.9378	Ala.	10/13/10	05/24/10	Yes	Yes
AL-9	301343087520200	BLM-2, Ala.	30.22881	-87.8672	Ala.	10/14/10	05/24/10	Yes	Yes
AL-10	301341087495200	Fort Morgan BLM-3, Ala.	30.22826	-87.8311	Ala.	10/14/10	05/24/10	Mix	Yes
LA-6	292708089521400	Bay Jimmy at Northeast Barataria Bay, La.	29.45222	-89.8706	La.	08/23/10	–	Yes	–
LA-22	294432090083100	Jean Lafitte National Park, La.	29.74222	-90.1419	La.	10/13/10	05/14/10	No	–

Table 1. U.S. Geological Survey pre-landfall and post-landfall sampling sites in the Gulf of Mexico, 2010 Deepwater Horizon oil spill.—Continued

[Sample dates are expressed as MM/DD/YY. **Abbreviations**: Ala, Alabama; BLM, Bureau of Land Management; dd, decimal degrees; Fla., Florida; La., Louisiana; M-1, Macondo-1 well; Miss., Mississippi; NWISweb, online National Water Information System (http://nwis.waterdata.usgs.gov/usa/nwis/qwdata); NWR, National Wildlife Refuge; St, Saint; Tex., Texas; USGS, U.S. Geological Survey; WSC, Water Science Center; —, no post-landfall sample was collected]

Map number	Station identifier	Site name	Latitude (dd)	Longitude (dd)	USGS WSC	Dates that site was sampled		M-1 oil in post-landfall[1]	
						Post-landfall	Pre-landfall[2]	Sediment	Tarball
LA-23	294406091511300	Cypremort Point, La.	29.735	-91.8536	La.	10/05/10	05/13/10	No	—
LA-24	292046090254500	Lake Felicity, La.	29.34611	-90.4292	La.	10/12/10	05/18/10	No	—
LA-25	293808092460200	Rockefeller Refuge Beach, La.	29.63556	-92.7672	La.	10/07/10	05/13/10	No	—
LA-26	291507090551800	Sister Lake, La.	29.25194	-90.9217	La.	10/08/10	05/17/10	No	—
LA-28	293424091321600	Point Chevreuil, La.	29.57333	-91.5378	La.	10/05/10	05/13/10	No	—
LA-29	294324089432500	Crooked Bayou, La.	29.72333	-89.7236	La.	10/13/10	05/18/10	No	—
LA-30	294108089234500	Mississippi River Gulf Outlet, La.	29.68556	-89.3958	La.	10/13/10	05/07/10	No	—
LA-31	291537089570100	Grand Isle Beach at State Park, La.	29.26028	-89.9503	La.	10/14/10	05/10/10	Yes[3]	Yes[3]
LA-32	291914089105500	Mississippi River at Main Pass, La.	29.32056	-89.1819	La.	10/07/10	05/07/10	No	—
LA-33	293518089364300	Breton Sound, La.	29.58833	-89.6119	La.	10/13/10	05/07/10	No	—
LA-34	300907089144500	Mississippi Sound at Grand Pass, La.	30.15194	-89.2458	La.	10/11/10	05/07/10	No	—
LA-35	285951089085600	Mississippi River at South Pass, La.	28.9975	-89.1489	La.	10/07/10	05/07/10	No	—
LA-36	285615089235600	Mississippi River at Southwest Pass, La.	28.9375	-89.3989	La.	10/14/10	05/07/10	Yes	—
LA-46	294456093394801	East Sabine, La.	29.74889	-93.6633	La.	10/06/10	05/10/10	No	—
MS-37	301309089044700	South Cat Island Beach, Miss.	30.21917	-89.0797	Miss.	10/14/10	05/07/10	Yes	Yes
MS-38	301227088582000	West Ship Island Beach, Miss.	30.2075	-88.9722	Miss.	10/14/10	05/07/10	No	—
MS-39	301358088533300	East Ship Island Beach, Miss.	30.23278	-88.8925	Miss.	10/11/10	05/07/10	No	Yes
MS-40	301425088440600	West Horn Island Beach, Miss.	30.24028	-88.735	Miss.	10/12/10	05/08/10	Yes	Yes
MS-41	301321088353300	East Horn Island Beach, Miss.	30.2225	-88.5925	Miss.	10/12/10	05/08/10	No	Yes
MS-42	301208088253600	Petit Bois Island Beach, Miss.	30.20222	-88.4267	Miss.	10/13/10	05/08/10	Yes	Yes
MS-43	301858089141000	Pass Christian Beach, Miss.	30.31611	-89.2361	Miss.	10/08/10	05/08/10	No	Yes
MS-44	302336088535800	Biloxi Beach, Miss.	30.39333	-88.8994	Miss.	10/07/10	05/08/10	No	No
MS-45	302034088325200	Pascagoula Beach, Miss.	30.34278	-88.5478	Miss.	10/14/10	05/08/10	No	No
TX-47	294057093572301	Texas Point, Tex.	29.6825	-93.9564	Tex.	10/06/10	05/10/10	No	—
TX-48	295542093521701	Sabine Lake, Tex.	29.92833	-93.8714	Tex.	—	05/10/10	—	—
TX-49	293324094220601	High Island, Tex.	29.55667	-94.3683	Tex.	10/07/10	05/10/10	No	—
TX-50	293429094332101	East Bay near Anahuac, Tex.	29.57472	-94.5558	Tex.	—	05/10/10	—	—
TX-51	291815094461001	Galveston Island, Tex.	29.30417	-94.7694	Tex.	10/13/10	05/10/10	No[3]	—
TX-52	294408094501101	Trinity Bay near Beach City, Tex.	29.73556	-94.8364	Tex.	—	05/11/10	—	—
TX-53	292318094430901	Bolivar Peninsula, Tex.	29.38833	-94.7192	Tex.	10/07/10	05/11/10	No	—
TX-54	292937094544001	Galveston Bay near Eagle Point, Tex.	29.49361	-94.9111	Tex.	—	05/11/10	No[3]	—
TX-55	291251094571401	West Bay, Galveston Island State Park, Tex.	29.21417	-94.9539	Tex.	10/14/10	05/11/10	No[3]	—
TX-56	290512095063101	San Luis Pass, Tex.	29.08667	-95.1086	Tex.	10/05/10	05/11/10	No	—

Water Samples

Water samples were collected to represent surf and suspended-sediment conditions at the time of sampling (Wilde and Skrobialowski, 2011). Samples were collected in wadable water about 60 to 90 centimeters (cm) deep by using the direct dip method. Samples were collected from depths of 15 to 30 cm below the surface, and at least 15 cm from the sea bottom to avoid collection of re-suspended bottom material. In general, sample containers were submerged to an appropriate depth, uncapped to fill the container to the appropriate volume, and recapped underwater. For analysis of trace and major elements and nutrients, water was collected in field-rinsed bottles, then poured into smaller sample bottles containing the appropriate chemical preservative (table 2). Bottles used for organic-contaminant samples were not field rinsed prior to sample collection to avoid over-representing oil in the water sample (Wilde and Skrobialowski, 2011). Quality-control (QC) samples collected for water included field (ambient) blanks, trip blanks, matrix spikes, and field replicates; these are described later in the report. Water samples were preserved, if appropriate, then stored on ice in coolers and shipped chilled at less than 4°C to the appropriate laboratory. Table 2 lists the laboratory, method code, sample containers, and preservatives for each class of analytes determined in water samples.

Sediment Samples

Wet-sediment core samples were collected from a 2-square meter or larger area at the land/water interface, or swash zone, on beaches and from bottom materials of streams that dissect wetland or marsh areas (Wilde and Skrobialowski, 2011). For post-landfall samples, samples were collected from an area and at a depth horizon to which oil could have penetrated (Wilde and Skrobialowski, 2011). Beach sediment samples were collected to a depth of 25 cm from the swash zone by using a Teflon scoop or core tube and were stored in a Teflon-lined bucket. Where possible, post-landfall sediment was collected at a comparable stage of tide as the pre-landfall sample collection at the same site. Marsh sediment samples were collected from a depth of 10 to 15 cm in submerged sediment by using a Teflon scoop and were stored in a Teflon-lined bucket. A single bulk-sediment sample was subdivided into subsamples for different analyses, including various chemical contaminants, percent moisture, pore-water toxicity, microorganisms, and oil-fingerprinting characteristics. The sub-samples for chemical analyses were chilled to less than 4°C and shipped to the appropriate laboratory (table 3).

Chemical Analyses

Water and sediment samples were analyzed for a variety of contaminants known to be associated with oil. Crude oil contains a complex mixture of many types of hydrocarbons, which range in size from 1 to 50 carbon atoms per molecule and in structure from simple, linear alkanes to branched or cyclic molecules (Georgia Coastal Research Council, 2010). These include polycyclic aromatic hydrocarbons (PAH), which are important because of their potential adverse effects on humans and aquatic life (U.S. Environmental Protection Agency, 2010, 2011a, and 2011b). Crude oil typically contains 1 to 2 percent PAHs, with the majority being alkylated PAHs (Operational Science Advisory Team, 2010). The M-1 well oil is a light, sweet oil with about 84 percent carbon, 4 percent hydrogen, and often less than 1 percent sulfur by weight (Georgia Coastal Research Council, 2010) and has an American Petroleum Institute (API) gravity of 38.8 degrees (Rosenbauer and others, 2010). "Light" indicates that the material has a low density due to the relatively high abundance of smaller, saturated alkane hydrocarbons. "Sweet" indicates there is little sulfur contamination (Georgia Coastal Research Council, 2010). The U.S. Environmental Protection Agency (2011a) identified nickel and vanadium as relevant to the Deepwater Horizon oil spill, and the general category of Louisiana sweet crude oil was reported to be low in trace elements, having 0.1 to 0.8 percent sulfur by weight, 0 to 4 milligrams per kilogram (mg/kg) vanadium, and 0 to 6 mg/kg nickel (Nadkarni, 1991). In two surveys of the general category of light crude oils, or those having an API gravity of 33 degrees or more, reported by the American Petroleum Institute (2011), maximum trace-element concentrations were less than 1 mg/kg for arsenic, barium, cadmium, chromium, cobalt, copper, mercury, molybdenum, lead, antimony, selenium, and thallium. Concentrations were higher for iron, nickel, tin, vanadium, and zinc, which had mean values in the 2 to 4 mg/kg range and maximum concentrations of 16 mg/kg for iron, 7 mg/kg for nickel, 10 mg/kg for tin, 20 mg/kg for vanadium, and 8 mg/kg for zinc. Concentrations of most trace elements are similar in different crude types, but nickel and vanadium tend to increase as crude oils become heavier (American Petroleum Institute, 2011).

In general, after oil is released into the environment, it is subject to various weathering processes, including dissolution, evaporation, emulsification, photo-oxidation, sedimentation, and biodegradation. The lower molecular-weight components tend to be lost through dissolution and evaporation, and photo-oxidation forms more water-soluble products, such as

Table 2. Chemical analysis of whole water samples from the Deepwater Horizon oil spill, Gulf of Mexico, 2010: laboratory, analytical methods, and sample treatment.

[**Abbreviations**: BTEX, benzene, toluene, ethylbenzene, xylenes and related compounds; CVAA, cold vapor atomic absorption; FA, filter-acidify; g, gram; GC, gas chromatography; H_2SO_4, sulfuric acid; HCl, hydrochloric acid; HNO_3, nitric acid; ICP, inductively coupled plasma; L, liter; LAB, laboratory; mL, milliliter; MS, mass spectrometry; Na, sodium; NH_4, ammonium; NWQL, USGS National Water Quality Laboratory; OCRL, Organic Carbon Research Laboratory; OES, optical emission spectrometry; oz, ounce; PAH, polycyclic aromatic hydrocarbon; poly, polypropylene plastic container; RA, raw acidified; SIM, selective ion monitoring; SVOC, semivolatile organic compound; TAL-CO, TestAmerica Laboratory, Arvada, Colorado; TAL-FL, TestAmerica Laboratory, Pensacola, Florida; TAL-VT, TestAmerica Laboratory, Burlington, Vermont; TKN, total Kjeldahl nitrogen; TP, total phosphorus; TPH-DRO, total petroleum hydrocarbons–diesel range organics; TPH-GRO, total petroleum hydrocarbons–gasoline range organics; USEPA, U.S. Environmental Protection Agency; USGS, U.S. Geological Survey; VOA, volatile organic analyses; VOC, volatile organic compounds; 1x, one of (or 2x means two of, etc.) the type of container that follows; C, degrees Celsius; <, less than; µm, micrometer]

Sample collected	Analytes	Lab	Method (code or schedule)	Container, preservation, and handling	Citations
Pre-landfall	PAH and other SVOC	USGS NWQL	GC/MS O-1316-87, O-3117-83, O-3118-83	1x 1-L baked, amber glass. Chill to 4°C without freezing	Fishman (1993); Wershaw and others (1987)
Pre- and post-landfall	PAH and other SVOC	TAL-FL	Continuous liquid/liquid extraction; GC/MS (8270D)	2x 1-L baked, amber glass. Leave headspace. Chill to 4°C without freezing. Ship overnight	USEPA (1986), Method SW-846-8270d
Pre- and post-landfall	Oil and grease	TAL-FL	n-Hexane extraction, gravimetry (1664A-HEM)	1x 1-L wide-mouth glass. Leave headspace. Preserve to pH 2 with HCl. Chill to 4°C without freezing. Ship overnight	USEPA (1999) Method 1664, revision A
Pre-landfall	VOC with BTEX	USGS NWQL	Purge and trap GC/MS O-5506-06 (2021)	3x 40-mL VOA septum vials. No bubbles. Chill to 4°C without freezing	Connor and others (1998); Wershaw and others (1987)
Post-landfall	VOC with BTEX	TAL-CO	Purge and trap, GC/MS (8260B)	2x 40-mL VOA septum vials. No bubbles. Chill to 4°C without freezing	USEPA (1986), Method SW-846-8260b
Pre- and post-landfall	VOC with BTEX	TAL-FL	Purge and trap, GC/MS (8260B)	2x 40-mL VOA septum vials. No bubbles. Chill to 4°C without freezing	USEPA (1986), Method SW-846-8260b
Pre-landfall	TPH-GRO with BTEX	TAL-CO	Purge and trap GC (8015B)	3x 40-mL VOA septum vials. No bubbles. Chill to 4°C	USEPA (1986), Method SW-846-8015b
Pre- and post-landfall	TPH-GRO	TAL-FL	Purge and trap GC (8015B)	2x 40-mL VOA septum vials. No bubbles. Chill to 4°C	USEPA (1986), Method SW-846-8015b
Post-landfall	TPH-DRO	TAL-FL	Flame ionization GC (8015C)	1x 1-L fired, amber glass. Leave headspace. Chill to 4°C without freezing	USEPA (1986), Method SW-846-8015c

Table 2.   Chemical analysis of whole water samples from the Deepwater Horizon oil spill, Gulf of Mexico, 2010: laboratory, analytical methods, and sample treatment.—Continued

[**Abbreviations**: BTEX, benzene, toluene, ethylbenzene, xylenes and related compounds; CVAA, cold vapor atomic absorption; FA, filter-acidify; g, gram; GC, gas chromatography; H_2SO_4, sulfuric acid; HCl, hydrochloric acid; HNO_3, nitric acid; ICP, inductively coupled plasma; L, liter; LAB, laboratory; mL, milliliter; MS, mass spectrometry; Na, sodium; NH_4, ammonium; NWQL, USGS National Water Quality Laboratory; OCRL, Organic Carbon Research Laboratory; OES, optical emission spectrometry; oz, ounce; PAH, polycyclic aromatic hydrocarbon; poly, polypropylene plastic container; RA, raw acidified; SIM, selective ion monitoring; SVOC, semivolatile organic compound; TAL-CO, TestAmerica Laboratory, Arvada, Colorado; TAL-FL, TestAmerica Laboratory, Pensacola, Florida; TAL-VT, TestAmerica Laboratory, Burlington, Vermont; TKN, total Kjeldahl nitrogen; TP, total phosphorus; TPH-DRO, total petroleum hydrocarbons–diesel range organics; TPH-GRO, total petroleum hydrocarbons–gasoline range organics; USEPA, U.S. Environmental Protection Agency; USGS, U.S. Geological Survey; VOA, volatile organic analyses; VOC, volatile organic compounds; 1x, one of (or 2x means two of, etc.) the type of container that follows; C, degrees Celsius; <, less than; μm, micrometer]

Sample collected	Analytes	Lab	Method (code or schedule)	Container, preservation, and handling	Citations
Pre-landfall	Trace elements, major ions	USGS NWQL	ICP-OES & ICP-MS (I-4471-97, I-4020-05, & I-4472-97) Method 2130	1x 125-mL poly, acid-rinsed (FA)	Garbarino and Stuzeski (1998); Garbarino (1999); Garbarino and others (2006)
Post-landfall	Trace elements, major ions	TAL-FL	ICP (6010B)	Use pre-cleaned, blanked RA bottle to collect dip sample and fill acidified 1x 250-mL poly bottle. Leave headspace. Preserve to pH<2 with HNO_3. Chill to 4°C without freezing. Ship overnight	USEPA (1986), Method SW-846-6010b
Post-landfall	Mercury	TAL-FL	CVAA (EPA 7470A)	Use pre-cleaned 250-mL glass bottle, acidify collected sample with 2 mL of 6 normal hydrochloric acid.	USEPA (1986), Method SW-846-7470a
Pre-landfall	Nutrients	USGS NWQL	Method IDs: I-4650-03, I-4522-85, I-4515-91, I-4610-01.	1x 125-mL poly. Add H_2SO_4. Chill to 4°C	Fishman and Friedman (1989); Patton and Kryskalla (2003); Patton and Truitt (1992, 2000)
Post-landfall	Nutrients	TAL-FL	Colorimetry (350.1, 365.1, 365.4)	Use pre-cleaned, blanked RA bottle to collect dip sample and fill acidified 1x 250-mL brown poly bottle. Leave headspace. Preserve to pH<2 with H_2SO_4. Chill to 4°C without freezing. Ship overnight	USEPA (2012)
Pre- and post-landfall	Dissolved organic carbon	USGS OCRL	Absorbance and fluorescence spectroscopy	1x 125-mL fired glass. Filter at laboratory. Chill to 4°C without freezing	Wilde and Skrobialowski (2011), appendix C-1

phenols, carboxylic acids, and ketones (Operational Science Advisory Team, 2011). Intermediate molecular-weight components can float and disperse in water, form emulsions, or sorb to sediment; the viscous, heavy components can form solid aggregates, or tarballs, that float or sink in water or sorb to sediment (American Petroleum Institute, 2003). Meanwhile, oil molecules are subject to microbial degradation at rates depending on the complexity of the oil molecules; degree of dispersion; environmental factors, such as temperature, oxygen, and nutrient concentrations; and the species and abundance of microbial organisms (Operational Science Advisory Team, 2011). The result is "weathered" crude oil that has a different composition from the oil originally released. A sample of weathered M-1 oil collected on April 27, 2010, was determined to contain aliphatic and cyclic hydrocarbons of, or greater than, C14—that is, with 14 or more carbon atoms. Benzene, toluene, ethylbenzene, xylene and related volatile (BTEX) compounds were not detected (State of Florida Oil Spill Academic Task Force, 2010). One goal of the nearshore sampling and chemical analysis was to characterize the weathering and shoreline degradation of the oil (Unified Area Command, 2010).

In the present study, contaminant classes determined in water and whole sediment included various organic compounds, trace and major elements, nutrients, and organic carbon. Trace and major elements, nutrients, and carbon also were analyzed in the fine sediment fraction, which is also called the silt-clay fraction, defined as less than 63-micrometer (μm) in size. As noted previously, there were changes in target analytes and analytical methods between the pre-landfall and post-landfall sampling periods. In September 2010, the Operational Science Advisory Team (2010, appendix F) recommended that future sample analyses in water and sediment include 43 PAH analytes, other organic compounds required for comparison to USEPA benchmarks for PAH mixtures, and metals. In the case of PAHs, this represented an expanded analyte list, and a change in the analytical method was made. For PAHs in sediment, pre-landfall samples were subsequently reanalyzed by using the updated analytical method. The analytical methods and laboratories that carried out the analyses are listed by analyte type in table 2 for water samples, and in table 3 for sediment samples, and are described briefly in the following sections. More detail is provided in the publications cited in tables 2 and 3. Analytical reporting levels are summarized in appendix 1.

Water

Organic contaminants analyzed in water included volatile organic compounds (VOC), PAHs and other semivolatile organic compounds (SVOC), dissolved organic carbon (DOC), gasoline-range organics having 6 to 10 carbon atoms, diesel-range organics having 10 to 28 carbon atoms,

and oil and grease (table 2). Most organic contaminants were determined by using gas chromatography with mass spectrometry (GC/MS). Most analyses were carried out at either the USGS National Water Quality Laboratory (NWQL) in Denver, Colorado, or the TestAmerica Laboratory in Pensacola, Florida (table 2), except for DOC, which was determined at the USGS Organic Carbon Research Laboratory (OCRL) in Boulder, Colorado.

Trace and major elements and nutrients were determined in water by various methods, including cold vapor atomic absorption spectrometry for mercury, and inductively coupled plasma-optical emission spectrometry (ICP-OES) or inductively coupled plasma-mass spectrometry (ICP-MS) for other trace elements (table 2). Analyses of water samples were carried out at either the USGS NWQL or the TestAmerica Laboratory in Florida (table 2). If water samples had high specific conductance (for example, greater than 2,000 microsiemens per centimeter) or high dissolved total solids, they were diluted prior to analysis by ICP-OES or ICP-MS methods for operational purposes and to approximate the matrices of the standards used to calibrate the instruments. High-salinity samples can cause an accumulation of solids in the sample-introduction system on ICP-OES and ICP-MS instruments, thereby compromising sensitivity (and therefore detection capability), accuracy, and precision (Tedmund M. Struzeski, Chemist, U.S. Geological Survey, Denver, Colo., written comm., Nov. 8, 2011).

References for the analytical methods that were used are cited in table 2, and individual analytes and their reporting levels in water are listed in appendix tables 1-1 and 1-3.

Sediment

Contaminants determined in whole, unsieved sediments included PAHs, alkylated PAH groups, other SVOCs, oil and grease, trace and major elements, nutrients, and carbon (table 3). Oil and grease in sediment was analyzed by the TestAmerica Laboratory in Arvada, Colorado, for pre-landfall samples and the TestAmerica Laboratory in Florida for post-landfall samples (table 3). PAHs in sediment were analyzed by GC/MS at the USGS NWQL for pre-landfall samples and the TestAmerica Laboratory in Florida for post-landfall samples. In addition, both pre-landfall and post-landfall samples were analyzed for PAHs and alkylated PAH groups at the TestAmerica Laboratory in Burlington, Vermont, by using GC/MS in the selective ion monitoring mode (SIM). Because the GC/MS SIM mode resulted in lower method detection limits (MDL), analytical results from the GC/MS SIM method were given precedence over results obtained by using GC/MS when both were available. Pre-landfall sediment samples were frozen for approximately 8 months prior to reanalysis in February 2011 by GC/MS SIM.

Table 3. Chemical analysis of sediment samples from the Deepwater Horizon oil spill, Gulf of Mexico, 2010: laboratory, analytical methods, and sample treatment.

[**Abbreviations**: AAS, atomic absorption spectrometry; Ag, silver; Cd, cadmium; CVAA, cold vapor atomic absorption; g, gram; GC, gas chromatography; HEM, hexane extractable material; Hg, mercury; ICP, inductively coupled plasma; L, liter; LAB, laboratory; MS, mass spectrometry; NWQL, USGS National Water Quality Laboratory; OES, optical emission spectrometry; oz., ounce; PAH, polycyclic aromatic hydrocarbon; Pb, lead; SCL, Sediment Chemistry Laboratory, Atlanta, Georgia; SIM, selective ion monitoring; SVOC, semivolatile organic compound; TAL-CO, TestAmerica Laboratory, Denver, CO; TAL-FL, TestAmerica Laboratory, Pensacola, Florida; TAL-VT, TestAmerica Laboratory, Burlington, Vermont; TC, total carbon; TIC, total inorganic carbon; TN, total nitrogen; TOC, total organic carbon; TP, total phosphorus; TS, total sulfur; USEPA, U.S. Environmental Protection Agency; USGS, U.S. Geological Survey; 1x, one of (or 2x means two of, etc.) the type of container that follows; C, degrees Celsius; <, less than; μm, micrometer]

Sample collected	Analysis	Lab	Method or schedule	Container	Citations
Pre-landfall	PAH and other SVOC	NWQL	GC/MS O-5506-06; schedules 5506, 5507	100 g. in 1x 1-L wide-mouth glass. Chill to 4°C	Zaugg and others (2006)
	PAH	TAL-VT	GC/MS SIM/isotope dilution (8270C SIM)	Reanalysis of frozen samples (above) sent to NWQL	USEPA (1986), Method SW-846-8270c
Post-landfall	PAH and other SVOC	TAL-FL	GC/MS (8270D)	1x 8-oz. wide-mouth glass. (Sample combined with oil & grease.) Chill to 4°C	USEPA (1986), Method SW-846-8270d
	PAH	TAL-VT	GC/MS SIM/isotope dilution (8270C SIM)	1x 4 oz. wide mouth glass. Chill to 4°C	USEPA (1986), Method SW-846-8270c
Pre-landfall	Oil & grease	TAL-CO	1664A-HEM	1x 8-oz. wide-mouth glass. (Sample combined with SVOC.) Chill to 4°C	USEPA (1999)
Post-landfall	Oil & grease	TAL-FL	1664A-HEM	1x 8-oz. wide-mouth glass. (Sample combined with SVOC.) Chill to 4°C	USEPA (1999)
Pre- and post-landfall	Trace and major elements, TN, TP, TS, TOC, and TIC	USGS SCL	Strong acid digestion; flame AAS (Ag, Cd, Pb), CVAA (Hg), or ICP-OES (other trace elements); combustion (TC, TN, TOC)	2x 18-oz. WhirlPak, whole sediments; 1x 18-oz. WhirlPak, <63 μm sediments. Keep away from light. Chill to 4°C without freezing	Horowitz and Stephens (2008); Wilde and Skrobialowski (2011), Appendix C-5

For most trace and major elements, whole sediment was subjected to strong acid digestion prior to chemical analyses at the USGS Sediment Chemistry Laboratory (SCL) in Atlanta, Georgia (table 3). This analysis generated total concentrations of trace and major elements, that is, 95 percent or more of the element present in sediment (Horowitz and Stephens, 2008). Silver, cadmium, and lead were determined by flame atomic absorption spectrometry, and other constituents were determined by ICP-OES. Mercury was digested separately and determined by cold vapor atomic absorption spectrometry. Total nitrogen, total carbon, and total organic carbon were determined by combustion.

Additional sediment subsamples were wet-sieved through a 63-μm polyester mesh to obtain the silt-clay fraction, which was subjected to the same strong acid-digestion procedure and analyzed for the same trace and major elements as whole sediment, for comparison to national baseline concentrations in fine sediment. Sieving sediment at 63 μm limits the grain-size effect, which results from finer material that typically contains higher trace-element concentrations than coarser material, and facilitates spatial and temporal comparisons (Horowitz and Stephens, 2008). The less than 63-μm fraction tended to have markedly lower sample mass than whole sediment. For about 20 samples, the less than 63-μm sample mass was insufficient to run a chemical analysis.

References for the analytical methods used are cited in table 3, and individual analytes and their reporting levels in sediment are listed in appendix tables 1-2, 1-4, and 1-5.

Quality-Control Samples

Three types of field QC samples were collected: blanks, replicates, and matrix spikes. The number of blanks, replicate sets, and matrix samples for laboratory spiking submitted to the various laboratories are shown in table 4.

Blanks

Blanks consist of samples prepared with water that is certified to be free of the analytes that will be measured by the laboratory. Blanks are used to estimate positive bias that can be caused by incidental contamination, which is the unintentional introduction of an analyte into the sample. For evaluation of potential contamination in water samples, three types of blanks were collected: field blanks, trip blanks, and equipment blanks. For evaluation of potential contamination in sediment samples, the only blanks collected were equipment blanks.

Table 4.   Number of blanks, matrix samples for laboratory spiking, and replicate sets from the 2010 Deepwater Horizon oil spill, submitted to U.S. Geological Survey and TestAmerica laboratories.

[**Abbreviations**: NWQL, National Water Quality Laboratory, Denver, Colorado; OCRL, Organic Carbon Research Laboratory, Boulder, Colorado; SCL, Sediment Chemistry Laboratory, Atlanta, Georgia; TAL-CO, TestAmerica Laboratory, Denver, Colorado; TAL-FL, TestAmerica Laboratory, Pensacola, Florida; TAL-VT, TestAmerica Laboratory, Burlington, Vermont; USGS, U.S. Geological Survey; –, no applicable samples of this type]

Laboratory	Blanks					Replicates				Matrix spikes			
	Pre-landfall	Post-landfall				Pre-landfall		Post-landfall		Pre-landfall		Post-landfall	
	Field	Equipment		Field	Trip	Water	Sediment	Water	Sediment	Water	Sediment	Water	Sediment
		Water	Sediment										
USGS NWQL	7	–	–	–	–	27	22	–	–	5	4	–	–
USGS SCL	–	–	–	–	–	–	9	–	9	–	–	–	–
USGS OCRL	5	6	4	–	–	9	–	3	–	–	–	–	–
TAL-CO	5	–	–	–	–	9	9	–	–	–	–	–	–
TAL-FL	–	5	3	4	31	–	–	7	7	–	–	5	3
TAL-VT	–	–	–	–	–	–	24	–	7	–	–	–	–

Field blanks were prepared by pouring blank water directly into sample bottles under ambient conditions at field sites. These are "field" blanks because they were prepared in the field by the same procedure used to collect environmental samples. In general, they did not contact any sampling equipment other than the sample bottles. In Wilde and Skrobialowski (2011), they are called "ambient" blanks because they were exposed to the atmosphere. Blanks and environmental samples for DOC and total nitrogen collected during the pre-landfall period were pumped from a collection bottle through a filter into a sample bottle. Field blanks enable the assessment of potential contamination of environmental water samples during sample preparation. Sources of contamination are not necessarily the same for pre-landfall and post-landfall samples, however, because (1) conditions could vary from one sampling period to the next and (2) pre-landfall samples and post-landfall samples were not analyzed at the same time and, in some cases, were analyzed by different laboratories (tables 2 and 4). Thus, pre-landfall field blanks should be compared only to pre-landfall water samples, and post-landfall field blanks compared only to post-landfall water samples.

Trip blanks were prepared at the TestAmerica Laboratory in Florida during the post-landfall period. These blanks were shipped to USGS offices, transported to field sites during sampling trips, and returned to the laboratory with environmental samples. Trip blanks generally are prepared only for VOCs and are used to evaluate whether environmental samples were contaminated during sample transport and analysis. Absence of detectable contaminants in a trip blank indicates there is no evidence that environmental samples were contaminated during transport and processing, but does not necessarily rule out contamination from other sources, such as ambient conditions at the site.

Equipment blanks were prepared in USGS offices by pumping blank water through water-sampling equipment, or by pouring blank water over sediment-sampling equipment, and collecting the rinsate in sample bottles. Blanks prepared by using water-sampling equipment are useful in this study only for comparison to environmental samples that were pumped through a filter—that is, pre-landfall samples analyzed for organic carbon and total nitrogen. Even for these analytes, the field blanks provide a more useful comparison than equipment blanks because they more completely represent potential sources of contamination. For sediment, equipment rinsate blanks were intended to indicate the potential for incidental contamination of environmental sediment samples from collection equipment and containers. Blank-water rinsate can be assumed to pick up contaminants that are removed easily from the sampling equipment, but it

might not represent certain processes, such as abrasion, that can occur during sediment-sample collection. Also, laboratory analysis of the blanks is done by using methods for water, rather than methods for sediment, and the potential sources of contamination during sample processing and laboratory analysis are not exactly the same for water as for sediment.

Data from blank samples can be used to estimate the potential for contamination in environmental samples in excess of concentrations that actually occur in the sampled matrix, which in this study is water or sediment. If a representative blank can be associated with each environmental sample, analytical results for the blanks can be used to qualify results for the environmental samples (U.S. Environmental Protection Agency, 1989, pages 16–17 in chapter 5). If the blank contains detectable levels of an analyte, concentrations of that analyte in the associated environmental samples should be censored unless they exceed five times the amount in the blank or, if the analyte is considered a common laboratory contaminant (such as acetone), ten times the amount in the blank.

Field Replicates

Field replicates are two or more environmental samples that are collected and prepared such that they are considered to be essentially identical in composition. Replicates are used to estimate variability of the analytical result. In this study, replicate water samples were collected sequentially by filling one set of sample bottles, followed immediately by filling a second, third, and, in some cases, a fourth set of bottles. Replicate sediment samples were collected by compositing a large amount of material in a single container. This material was homogenized, and replicate subsamples were scooped into separate sample containers.

Statistical evaluation of replicate variability is based on the standard deviation of measured values in the primary environmental sample and the replicate sample, or samples. For many analytes, variability is correlated with the mean concentration of that analyte in the replicates (Martin, 2002; Mueller and Titus, 2005). Within a range of low concentrations, standard deviation of replicates generally is uniform, but at higher concentrations, standard deviation tends to increase in proportion to concentration. Within this higher range, the relative standard deviation (RSD), defined as the standard deviation of replicate results divided by the mean concentration, is generally uniform. Therefore, over the low-concentration range, variability is estimated as the average standard deviation of replicates; over the high-concentration range, variability is estimated as the average RSD.

Typically, replicate variability is similar to the analytic error of laboratory methods, having RSDs ranging from a few percent to around 10 percent. If variability is substantially higher than this range, it could interfere with certain types of data interpretation. For example, high variability adds uncertainty when comparing data to a standard or benchmark. Also, it can reduce the likelihood of finding statistically significant differences in comparisons among groups of data over time or space. Variability is less likely to affect the central tendency (for example, the mean and median) of data distributions, but can increase the spread and range.

Matrix Spikes

Matrix spikes are samples fortified, or "spiked," with known concentrations of analytes that will be measured by the laboratory. Spikes are used to estimate positive or negative bias in the analytical result caused by matrix effects—that is, chemical, physical, or biological characteristics of the sample material (water or sediment itself) that can interfere with chemical analysis of the sample. Matrix spike samples were collected in the same manner as field replicates; subsequently, these samples were spiked in the laboratory to introduce a known amount of the analytes of interest.

Method performance is determined by spike recovery, which is the measured amount of analyte expressed as a percentage of the known spiked amount. Recovery is calculated from analyte concentrations in the spiked sample compared to a replicate environmental sample that was not spiked. Recovery can be poorly estimated if the analyte concentration in the background environmental sample is similar to, or greater than, the expected concentration of the spiked addition.

Generally, recovery is within a few percent of 100 for analytes that are not affected by method or sample-matrix interferences, though the acceptable range can extend to within 10 to 20 percent for some analytes. Poor recovery is more typically low, rather than high. For constituents with chronically poor recovery, some aspects of data interpretation require qualification; for example, the detection frequency and the likelihood of exceeding a standard or benchmark can be underestimated.

Spikes are used most often for organic compounds because the analytical methods involve extraction and analysis steps that can be affected by other chemicals in the sample. For example, naturally occurring organic matter can be co-extracted with anthropogenic organic compounds in a sample and interfere with GC analysis.

Laboratory Quality-Control Procedures

Laboratory QC procedures include analysis of reagent blanks (also called method blanks), spikes, standard reference materials, and surrogate compounds. Each laboratory has its own QC procedures and analyses in order to assess the quality of the data and meet performance standards. It is beyond the scope of this report to describe the laboratory QC sampling, except in one regard—where contamination was detected in reagent blanks, this information was considered in data analysis for this report. Laboratory reagent blanks are processed and analyzed along with each set of environmental samples and are used to monitor for incidental contamination introduced during sample processing and analysis at the laboratory.

Water- and Sediment-Quality Benchmarks

Concentrations of trace and major elements and organic contaminants were compared to various benchmarks to assess the potential for adverse effects on human health or aquatic life. Benchmark comparisons were made for all available samples, including field replicate samples and samples from sites sampled in only one study period, to maximize the information gained from the dataset regarding benchmark exceedance at the sampled sites.

Contaminant concentrations in water were compared to benchmarks for protection of human health and aquatic life, whereas concentrations in whole sediment were compared to sediment-quality benchmarks for protection of benthic organisms. The benchmarks used were those recommended by the U.S. Environmental Protection Agency (2010, 2011a, and 2011b) on its web site, "EPA Response to BP Spill in the Gulf of Mexico, Coastal Water Sampling" (U.S. Environmental Protection Agency, 2011c), supplemented by screening-level benchmarks from the NOAA Office of Response and Restoration (Buchman, 2008). In addition, trace- and major-element and nutrient concentrations in the less than 63-μm sediment fraction were compared to national baseline concentrations in bed sediments of U.S. rivers from Horowitz and Stephens (2008).

Benchmark values are listed for organic contaminants in table 5 and for trace and major elements in table 6. Table 6D lists baseline concentrations for trace and major elements and nutrients in the less than 63-μm sediment fraction.

Human-Health Benchmarks for Water

Human-health benchmarks are based on potential cancer and non-cancer risks associated with recreational exposure to oil-contaminated water. They were developed by U.S. Environmental Protection Agency (2010) in coordination with the U.S. Department of Health and Human Services. These benchmarks consider both skin contact and incidental ingestion of water by a child swimmer, assuming 90 hours of exposure. Human-health benchmark values are available for five VOCs, six PAHs, and two trace elements—nickel and vanadium—in water (tables 5C and 6B).

Aquatic-Life Benchmarks for Water

For water samples, potential toxicity to aquatic life was assessed by comparison to two types of benchmarks: (1) a toxic-unit (TU) approach for mixtures of PAHs and BTEX compounds (table 5A) and (2) marine benchmarks for individual contaminants (tables 5B and 6A).

Toxic-Unit Benchmarks for PAH-BTEX Mixtures in Water

Because PAHs and BTEX compounds share a common mechanism of action, toxicity is expected to be additive. A toxic-unit approach is used, therefore, in which the concentration of each component (i) of the mixture is divided by a potency factor to determine its toxic-unit concentration (TU_i). The TU_i values for all components in the mixture are summed and the sum ($\sum TU_i$) is compared to a hazard index of 1 (U.S. Environmental Protection Agency, 2011a). Separate TU computations are made for acute and chronic toxicity by using acute and chronic potency factors, respectively (table 5A). Because alkylated PAHs (for example, C1- and C2-alkylated naphthalenes) tend to have comparable or greater toxicity to aquatic life than parent PAHs (for example, naphthalene itself), it is important to include alkylated PAHs in TU calculations. Because alkylated PAHs were not measured in water in this study, concentrations of alkylated PAHs were estimated from parent PAHs by using appropriate alkylation multipliers, as recommended by USEPA (Mount, 2010). USEPA developed these multipliers by using the analysis of a tarball that was collected at Dauphin Island during the current oil spill and checked them against oil composition data from other sources, including the Exxon Valdez oil. Because the present study did not analyze 2 of the 18 parent PAHs that should be included in the $\sum TU_i$ value for water—benzo(e) pyrene and perylene—these two compounds were omitted from $\sum TU_i$ calculations, which therefore could be biased low.

An acute or chronic $\sum TU_i$ value greater than 1 indicates that the sample has the potential to cause an acute or chronic effect, respectively, on aquatic organisms such as fish, crabs, and clams. The PAHs and BTEX contaminants included in the $\sum TU_i$ calculations for water are provided in table 5A, along with their acute and chronic potency factors and alkylation multipliers.

Marine Benchmarks

Marine benchmarks for acute or chronic exposure to individual contaminants are available from various sources for many organic contaminants (table 5B) and trace elements (table 6A). These values were obtained from Buchman (2008), who compiled acute and chronic marine benchmarks from multiple sources. Most values were from the USEPA, such as ambient water-quality criteria and Tier II Species-Acute Values, which were supplemented by benchmarks from Canada, British Columbia, and New Zealand. As such, individual benchmarks were not necessarily derived the same way, and exceedance of one benchmark can mean something slightly different from exceedance of another, as shown in these examples:

- The USEPA chronic water-quality criterion is the highest concentration of a pollutant that aquatic organisms can be exposed to for an extended period without deleterious effects. The acute water-quality criterion is the highest concentration that aquatic organisms can be exposed to for a short period (1-hour average) without deleterious effects. Both are intended to protect 95 percent of a diverse group of genera and should not be exceeded more than once every 3 years.

- Canadian aquatic-life guidelines are based on toxicity data for the most sensitive species of plants and animals found in Canadian waters; they are intended to protect all forms of aquatic life during all stages of the aquatic life cycles and should not be exceeded at any time (Canadian Council of Ministers of the Environment, 1999).

- The British Columbia guidelines set safe conditions or levels that have province-wide application and are designed to protect marine aquatic life (British Columbia Ministry of Environment, 2010). They are intended to protect all forms of aquatic life and all stages of their life cycle during indefinite exposure (Meays, 2010). If a single guideline is recommended, it represents a long-term no-effect level and should not be exceeded at any time. For some substances, both maximum (acute) and average (chronic) guidelines are recommended; acute guidelines apply in the initial dilution zone, and chronic guidelines apply everywhere else (Meays, 2010). In addition, British Columbia has working guidelines for additional contaminants that were obtained from various North American jurisdictions, but have not yet been fully assessed by the Ministry of Environment; they represent the best guidance the Ministry can provide about safe levels of these substances in the environment (Nagpal and others, 2006).

- Trigger values from New Zealand are derived by fitting an appropriate statistical distribution to the no-observed-effect-concentration data available for a given contaminant, and estimating a concentration that protects 95 percent of species in the environment (Australian and New Zealand Environment and Conservation Council, 2000).

Table 5A. Benchmark values for organic contaminants: toxic-unit benchmarks ($\sum TU_i$) for PAH and BTEX compounds in water.

[**Abbreviations**: BP, British Petroleum; BTEX, benzene, toluene, ethylbenzene, xylenes and related compounds; CASRN, Chemical Abstracts Service Registry Number; PAH, polycyclic aromatic hydrocarbons; TU, toxic unit; μg/L, microgram per liter; >, greater than]

Organic contaminant	CASRN	Acute divisor[1] (µg/L)	Chronic divisor[1] (µg/L)	Multiplier[2]	Analyzed in this study
Benzene	71-43-2	27,000	5,300	1	Yes
Cyclohexane	110-82-7	1,900	374	1	Yes
Ethylbenzene	100-41-4	4,020	790	1	Yes
Isopropylbenzene (cumene)	98-82-8	2,140	420	1	Yes
Total xylene (*o, m* and/or *p*)	108-38-3	3,560	700	1	Yes
Methylcyclohexane	108-87-2	463	91.0	1	Yes
Toluene	108-88-3	8,140	1,600	1	Yes
Naphthalene	91-20-3	803	193	120	Yes
Acenaphthylene	208-96-8	1,280	307	1	Yes
Acenaphthene	83-32-9	232	55.8	1	Yes
Fluorene	86-73-7	164	39.3	14	Yes
Phenanthrene	85-01-8	79.7	19.1	6.8	Yes
Anthracene	120-12-7	86.1	20.7	1	Yes
Fluoranthene	206-44-0	29.6	7.11	1	Yes
Pyrene	129-00-0	42.0	10.1	2.1	Yes
Benz[a]anthracene	56-55-3	9.28	2.23	1	Yes
Chrysene	218-01-9	8.49	2.04	5	Yes
Perylene	198-55-0	3.75	0.901	1	No
Benzo[b]fluoranthene	205-99-2	2.82	0.677	1	Yes
Benzo[k]fluoranthene	207-08-9	2.67	0.642	1	Yes
Benzo[e]pyrene	192-97-2	3.75	0.901	1	No
Benzo[a]pyrene	50-32-8	3.98	0.957	1	Yes
Indeno[1,2,3-cd]pyrene	193-39-5	1.14	0.275	1	Yes
Dibenz[a,h]- anthracene	53-70-3	1.17	0.282	1	Yes
Benzo[g,h,i]perylene	191-24-2	1.83	0.439	1	Yes

[1]The Toxic Unit ($\sum TU_i$) benchmark is computed by dividing the concentration of each individual compound by its potency divisor (acute or chronic), then adding the ratios for all compounds in the sample to calculate the combined toxicity. A $\sum TU_i$ benchmark value >1 indicates an exceedance.

[2]Because alkylated PAHs were not analyzed in water, their concentrations were estimated by applying a multiplier to the parent PAH concentration.

Table 5B. Benchmark values for organic contaminants: supplemental aquatic-life benchmarks for organic contaminants in water.

[**Abbreviations**: ANZ, Australian and New Zealand guideline trigger value (Australian and New Zealand Environment and Conservation Council, 2000); BC, British Columbia water-quality guideline for marine aquatic life (British Columbia Ministry of Environment, 2010; Nagpal and others, 2006); C, Value for chemical class; CA, Canadian water-quality guideline for the protection of marine aquatic life (Canadian Council of Ministers of the Environment, 2011); Eco, Ecotox threshold (U.S. Environmental Protection Agency [USEPA], 1996); LOEL, USEPA LOEL value (unverified) from Buchman (2008), who compiled LOELs previously published by USEPA; MW, molecular weight; p, proposed value (unverified) from Buchman (2008); PAH, polycyclic aromatic hydrocarbons; S, value for summation of isomers; #, chronic value derived by division of acute value by 10; <, less than; –, no benchmark]

Contaminant	Marine acute (µg/L)[1,2]	Marine chronic (µg/L)[1,2]	Contaminant	Marine acute (µg/L)[1,2]	Marine chronic (µg/L)[1,2]
1,1,1-Trichloroethane	31,200 LOEL	3,120 #	Dibromomethane (methylene bromide)	12,000 C, LOEL	6,400 C, LOEL
1,1,2,2-Tetrachloroethane	9,020 LOEL	902 #	Dichlorobenzenes	1,970 S, LOEL	–
1,1,2-Trichloroethane	–	1,900 ANZ	Dichloromethane (methylene chloride)	12,000 C, LOEL	6,400 C, LOEL
1,1-Dichloroethene	224,000 S, LOEL	–			
1,2,4-Trichlorobenzene	160 C, LOEL	5.4 CA	Diethyl phthalate	2,944 C, LOEL	3.4 C, LOEL
1,2-Dichlorobenzene	<1,970 S, LOEL	42 CA	Dimethyl phthalate	2,944 C, LOEL	3.4 C, LOEL
1,2-Dichloroethane	113,000 LOEL	11,300 #	Di-n-butyl phthalate	2,944 C, LOEL	3.4 C, LOEL
1,2-Dichloroethene (cis or trans)	224,000 S, LOEL	–	Di-n-octyl phthalate	2,944 C, LOEL	3.4 C, LOEL
1,2-Dichloropropane	10,300 S, LOEL	3,040 S, LOEL	Ethylbenzene	430 LOEL	25 CA
1,3-Dichlorobenzene	<1,970 S, LOEL	–	Fluoranthene	40 C, LOEL	11 Eco
1,3-Dichloropropene (cis or trans)	790 S, LOEL	–	Fluorene	300 C, LOEL	–
			Hexachlorobenzene	160 C, LOEL	129 C, LOEL
1,4-Dichlorobenzene	<1,970 S, LOEL	129 C, LOEL	Hexachlorobutadiene	32 LOEL	3.2 #
2,4,5-Trichlorophenol	240 p	11 p	Hexachlorocyclopentadiene	7 LOEL	0.7 #
2,4-Dinitrophenol	4,850 C, LOEL	–	Hexachloroethane	940 LOEL	94 #
2,4-Dinitrotoluene	590 S, LOEL	370 S, LOEL	Indeno[1,2,3-cd]pyrene	300 C, LOEL	–
2-Chloronaphthalene	7.5 C, LOEL	–	Isophorone	12,900 LOEL	1,290 #
2-Methylnaphthalene	300 C, LOEL	–	Methyl tert-butyl ether (MTBE)	–	5,000 CA
4-Chloroaniline	160 C, LOEL	129 C, LOEL			
4-Nitrophenol	4,850 C, LOEL	–	Monochlorobenzene	160 C, LOEL	25 CA
Acenaphthene	970 LOEL	40 Eco	Naphthalene	2,350 LOEL	1.4 CA
Acenaphthylene	300 C, LOEL	–	Nitrobenzene	6,680 LOEL	668 #
Anthracene	300 C, LOEL	–	N-Nitrosodiphenylamine	3,300,000 C, LOEL	–
Benzene	5,100 LOEL	110 CA	PAHs, high MW	300 C, LOEL	–
Benzo(a)anthracene	300 C, LOEL	–	PAHs, low MW	300 C, LOEL	–
Benzo(a)pyrene	300 C, LOEL	–	PAHs, total	300 C, LOEL	–
Benzo(b)fluoranthene	300 C, LOEL	–	Pentachlorophenol	13	7.9
Benzo(ghi)perylene	300 C, LOEL	–	Phenanthrene	7.7 p	4.6 p
Benzo(k)fluoranthene	300 C, LOEL	–	Sum dichloroethenes	224,000 S, LOEL	–
Benzyl n-butyl phthalate	2,944 C, LOEL	3.4 C, LOEL	Tetrachloroethene (PCE)	10,200 LOEL	450 LOEL
Bis(2-chloroethoxy)methane	12,000 C, LOEL	6,400 C, LOEL	Tetrachloromethane	50,000 LOEL	5,000 #
Bis(2-ethylhexyl) phthalate	400 p	360 p	Toluene	6,300 LOEL	215 CA
Bromodichloromethane	12,000 C, LOEL	6,400 C, LOEL	trans-1,2-Dichloroethene	224,000 S, LOEL	–
Chrysene	300 C, LOEL	–	trans-1,3-Dichloropropene	790 S, LOEL	–
cis-1,2-Dichloroethene	224,000 S, LOEL	–	Trichlorobenzenes	160 C, LOEL	<5.4 CA
cis-1,3-Dichloropropene	790 S, LOEL	–	Trichloroethene (TCE)	2,000 LOEL	200 #
Dibenzo(ah)anthracene	300 C, LOEL	–	Trichlorofluoromethane	12,000 C, LOEL	6,400 C, LOEL
Dibromochloromethane	12,000 C, LOEL	6,400 C, LOEL			

Table 5C. Benchmark values for organic contaminants: human-health benchmarks (recreational contact) for organic contaminants in water.

[**Abbreviations**: C, cancer endpoint; CASRN, Chemical Abstracts Service Registry Number; HH, human health; NC, noncancer effects endpoint; µg/L, microgram per liter]

Organic contaminant	CASRN	HH benchmark (child swimmer) (µg/L)	Cancer/ noncancer
Benzene	71-43-2	380	C
Isopropylbenzene (cumene)	98-82-8	20,000	NC
Ethylbenzene	100-41-4	610	C
Total xylene[1]	108-38-3	18,000	NC
Toluene	108-88-3	120,000	NC
2-Methylnaphthalene	91-57-6	170	NC
Naphthalene	91-20-3	1,800	NC
Acenaphthene	83-32-9	2,500	NC
Fluorene	86-73-7	12,000	NC
Anthracene	120-12-7	22,000	NC
Pyrene	129-00-0	4,100	NC

[1]Analyzed in this study as total xylene in some samples, and as the summed concentrations of *ortho, meta*, and *para* isomers in other samples.

Sediment-Quality Benchmarks

Potential effects of sediment contaminants on benthic organisms were assessed by comparing contaminant concentrations to benchmarks derived by using two different approaches: equilibrium partitioning and empirical biological-effects correlation. In the equilibrium-partitioning approach, an equilibrium-partition coefficient (K_{oc}) is used to calculate the contaminant concentration in sediment that corresponds to the concentration in interstitial water, or pore water, above which toxic effects on aquatic organisms could occur (Di Toro and others, 1991). This approach assumes that contaminants are in equilibrium between water and sediment organic carbon, and postulates a theoretical causal relation between chemical bioavailability and chemical toxicity in different sediments. Equilibrium-partitioning (EqP) benchmarks are available for nonionic-organic contaminants, including PAH mixtures and some individual organic contaminants, and are described later in this section.

In contrast, the biological-effects correlation approach consists of matching sediment-chemistry measurements with biological-effects measurements to relate the incidence of biological effects in field sediments to the concentration of an individual contaminant at a particular site. The matching measurements come primarily from field studies, and sometimes from spiked sediment bioassays. The dataset of matching measurements is used to identify a level of concern for an individual contaminant that is associated with a certain probability of observing adverse effects on benthic organisms in studies where that contaminant was measured. This approach is empirically based and does not indicate a direct cause-and-effect relation between chemical contamination and biological effects. It assumes that the contaminant measured is responsible for the effects observed, although field sediment samples typically contain complex mixtures of chemical contaminants (see, for example, MacDonald and others, 1996; Burgess and others, 2003; Hyland and others, 2003). Empirical, or correlative, benchmarks for both organic contaminants and trace elements are available from a number of sources, which are described later in this section.

Equilibrium-Partitioning Sediment Benchmarks for PAH Mixtures

As in water, toxicity to PAHs and BTEX compounds in sediment is expected to be additive. The bioavailability of nonionic organic compounds in sediment, however, is assumed to be controlled by sorption to sediment organic carbon. Therefore, the toxic unit approach in sediment first requires that measured concentrations of the contaminants be normalized to the total organic carbon (TOC) content of the sediment. Then, the TOC-normalized concentration of each component compound (i) is divided by its potency factor to obtain its equilibrium-partitioning sediment benchmark toxic-unit concentration ($ESBTU_i$), and the $ESBTU_i$ values are summed for all components in the sediment mixture to obtain the equilibrium-partitioning sediment benchmark toxic units ($\sum ESBTU_i$) for that sediment sample (U.S. Environmental Protection Agency, 2011b). Separate calculations are made for acute and chronic exposure by using acute and chronic potency factors. The PAHs included in $\sum ESBTU_i$ calculations consist of both parent PAHs and alkylated PAHs because the latter have comparable, or greater, toxicity than the former (table 5D). Just as in the TU procedure for water, the ESBTU procedure for sediment calls for using alkylation multipliers if data for alkylated PAH groups are not available. In this study, data were available for alkylated PAHs in all sediment samples, so alkylation multipliers were not used. BTEX compounds were not determined in sediment, however, so calculated $\sum ESBTU_i$ values could be slightly low; this bias is expected to be minimal in shoreline sediments because BTEX compounds are volatile, were not detected in weathered M-1 oil (State of Florida Oil Spill Academic Task Force, 2010), and are not expected to persist in sediment (Mount, 2010).

An acute or chronic $\sum ESBTU_i$ value greater than 1 indicates that the sample has the potential to cause an acute or chronic effect, respectively, on sediment-dwelling organisms, such as crabs, clams, and worms. The contaminants included in the ESBTU calculations, and their potency factors and multipliers, are provided in table 5D.

Table 5D.   Benchmark values for organic contaminants: equilibrium-partitioning sediment benchmark toxic units ($\Sigma ESBTU_i$) for PAH and BTEX compounds in sediment.

[**Abbreviations**: BP, British Petroleum; BTEX, benzene, toluene, ethylbenzene, xylenes and related compounds; CASRN, Chemical Abstracts Service Registry Number; $\Sigma ESBTU_i$, equilibrium-partitioning sediment benchmark toxic unit; PAH, polycyclic aromatic hydrocarbons; µg/kg-oc, microgram per kilogram of sediment organic carbon; >, greater than; –, not applicable]

Organic contaminant	CASRN	Acute divisor[1] (µg/kg-oc)	Chronic divisor[1] (µg/kg-oc)	Multiplier[2]	Analyzed in this study
Benzene	71-43-2	3,360,000	660,000	1	No
Cyclohexane	110-82-7	4,000,000	786,000	1	No
Ethylbenzene	100-41-4	4,930,000	970,000	1	No
Isopropylbenzene (cumene)	98-82-8	5,750,000	1,130,000	1	No
Total xylene	108-38-3	4,980,000	980,000	1	No
Methylcyclohexane	108-87-2	4,960,000	976,000	1	No
Toluene	108-88-3	4,120,000	810,000	1	No
Naphthalene	91-20-3	1,600,000	385,000	120	Yes
C1-Naphthalenes	–	1,850,000	444,000	–	Yes
C2-Naphthalenes	–	2,120,000	510,000	–	Yes
C3-Naphthalenes	–	2,420,000	581,000	–	Yes
C4-Naphthalenes	–	2,730,000	657,000	–	Yes
Acenaphthylene	208-96-8	1,880,000	452,000	1	Yes
Acenaphthene	83-32-9	2,040,000	491,000	1	Yes
Fluorene	86-73-7	2,240,000	538,000	14	Yes
C1-Fluorenes	–	2,540,000	611,000	–	Yes
C2-Fluorenes	–	2,850,000	686,000	–	Yes
C3-Fluorenes	–	3,200,000	769,000	–	Yes
Phenanthrene	85-01-8	2,480,000	596,000	6.8	Yes
Anthracene	120-12-7	2,470,000	594,000	1	Yes
C1-Phenanthrenes/anthracenes	–	2,790,000	670,000	–	Yes
C2-Phenanthrenes/anthracenes	–	3,100,000	746,000	–	Yes
C3-Phenanthrenes/anthracenes	–	3,450,000	829,000	–	Yes
C4-Phenanthrenes/anthracenes	–	3,790,000	912,000	–	Yes
Fluoranthene	206-44-0	2,940,000	707,000	1	Yes
Pyrene	129-00-0	2,900,000	697,000	2.1	Yes
C1-pyrene/fluoranthenes	–	3,200,000	770,000	–	Yes
Benz(a)anthracene	56-55-3	3,500,000	841,000	1	Yes
Chrysene	218-01-9	3,510,000	844,000	5	Yes
C1-Chrysenes/benzanthracenes	–	3,870,000	929,000	–	Yes
C2-Chrysenes/benzanthracenes	–	4,200,000	1,010,000	–	Yes
C3-Chrysenes/benzanthracenes	–	4,620,000	1,110,000	–	Yes
C4-Chrysenes/benzanthracenes	–	5,030,000	1,210,000	–	Yes
Perylene	198-55-0	4,020,000	967,000	1	Yes
Benzo(b)fluoranthene	205-99-2	4,070,000	979,000	1	Yes
Benzo(k)fluoranthene	207-08-9	4,080,000	981,000	1	Yes
Benzo(e)pyrene	192-97-2	4,020,000	967,000	1	Yes
Benzo(a)pyrene	50-32-8	4,020,000	965,000	1	Yes
Indeno(1,2,3-cd)pyrene	193-39-5	4,620,000	1,110,000	1	Yes
Dibenz(a,h) anthracene	53-70-3	4,660,000	1,120,000	1	Yes
Benzo(g,h,i)perylene	191-24-2	4,540,000	1,090,000	1	Yes

[1]The $\Sigma ESBTU_i$ benchmark is computed by dividing the sediment organic carbon-normalized concentration of each individual compound by its potency divisor (acute or chronic), then adding the ratios for all compounds to calculate the combined toxicity. An $\Sigma ESBTU_i$ benchmark value >1 indicates an exceedance.

[2]For samples with no data available for alkylated PAHs, concentrations were estimated by applying a multiplier to the parent PAH concentration.

Equilibrium-Partitioning Sediment Benchmarks for Individual Contaminants

As with the ESBTU approach described previously for PAH-BTEX mixtures, these EqP benchmarks are based on equilibrium-partitioning theory, but they apply to individual contaminants rather than contaminant mixtures. The acute and chronic EqP benchmarks are in units of microgram per gram (μg/g) of sediment TOC, so that measured contaminant concentrations in dry weights must be normalized to sediment TOC prior to comparison with these benchmarks (table 5E). Acute and chronic EqP benchmarks are based on acute or chronic toxicity to aquatic life, respectively, and represent the concentration of chemicals in sediment that are predictive of biological effects, protective of the presence of benthic organisms, and applicable to the range of natural sediments from lakes, streams, estuaries, and near-coastal marine waters (U.S. Environmental Protection Agency, 2008). Exceedance of an individual EqP benchmark indicates that effects can occur if the contaminant in question is bioavailable as predicted by EqP theory; in general, the degree of effect that is expected increases with increasing exceedance of the benchmark (U.S. Environmental Protection Agency, 2008). Individual EqP benchmark values are listed in table 5E.

Empirical Sediment Benchmarks

Several types of empirical benchmarks have been developed on the basis of correlations between measured chemical concentrations and observed toxicity in field sediments. As such, they define concentrations in sediments that are associated with certain types and levels of toxicity. These benchmarks typically come in pairs: lower screening values define concentrations below which adverse effects are not expected and upper screening values define concentrations above which adverse effects are likely or frequent. Four such pairs of sediment benchmarks are listed; benchmark types and values are shown in tables 5E and 6C. In this study, two supplementary benchmarks—Washington State's apparent effect threshold (AET; tables 5E and 6C) and the USEPA's EqP benchmark (table 5E)—are grouped with upper screening values because they indicate concentrations above which toxicity is likely.

- **Apparent Effect Threshold.** These values are based on matching sediment chemistry and toxicity data from Puget Sound. The AET is the concentration of an individual contaminant above which a particular adverse biological effect is always expected (Barrick and others, 1988). Different types of AETs represent different indicators of toxicity, including amphipod mortality, benthic abundance, Microtox, and oyster larval development. For a given contaminant, the AET value shown in table 5E or 6C represents the lowest available AET value, as determined by Buchman (2008). Because of its definition, the AET was considered an upper screening value in this study.

- **Effects-Range Low and Effects-Range Median.** These were derived from matching sediment chemistry and toxicity data. The effects range-low (ERL) corresponds to the lower 10th percentile of the matched data for a given contaminant and represents the contaminant concentration below which effects are rarely observed. The effects range-median (ERM) corresponds to the 50th percentile of the matched data and represents the contaminant concentration above which adverse effects frequently occur (Long and Morgan, 1991).

- **Threshold Effect Level and Probable Effect Level.** The Canadian threshold effect level (TEL) defines a concentration below which adverse effects are rarely anticipated and above which adverse effects are occasionally anticipated, whereas the probable effect level (PEL) defines a concentration above which adverse effects are frequently anticipated. Both the Canadian TEL and PEL are empirically based and were derived by compiling data from multiple types of studies in the literature, including equilibrium partitioning studies, guidelines from other jurisdictions, spiked-sediment toxicity tests, and field studies from throughout North America (Canadian Council of Ministers of the Environment, 1995 and 2001). The TEL and PEL values for a given contaminant were selected so that fewer than 25 percent of adverse effects occur below the TEL and more than 50 percent of adverse effects occur above the PEL (Canadian Council of Ministers of the Environment, 2001).

- **Threshold Effect Concentration and Probable Effect Concentration.** The consensus-based threshold effect concentration (TEC) from MacDonald and others (2000) defines the concentration below which adverse effects on sediment-dwelling organisms are not expected to occur. The consensus-based probable effect concentration (PEC) defines the concentration of sediment-associated contaminants above which adverse effects on sediment-dwelling organisms are likely to be observed. These guidelines were developed by compiling multiple sediment-quality guidelines for a given contaminant, including both causally and empirically based guidelines, identifying those that meet certain selection criteria, and selecting the geometric mean as the consensus-based guideline.

- **T20 and T50.** These were derived from logistic regression models that predict the probability of toxicity to marine amphipods by using a large database of matching sediment chemistry and toxicity data representing coastal North America (Field and others, 2002). The T20 and T50 for an individual contaminant consist of concentrations of that contaminant that are associated with a 20 percent or 50 percent probability, respectively, of observing toxicity.

Table 5E. Benchmark values for organic contaminants: supplemental aquatic-life benchmarks for organic contaminants in sediment.

[Abbreviations: AET, apparent effects threshold; B, bivalve; E, echinoderm larvae; EqP, equilibrium partitioning sediment benchmark; ERL, effects range-low; ERM, effects range-median; I, infaunal community index; L, larvae; M, microtox assay; N, neanthes bioassay; NOAA, National Oceanic and Atmospheric Administration; O, oyster larvae; PEC, probable effect concentration; PEL, probable effect level; SV, screening value; T20, chemical concentration corresponding to 20 percent probability of toxicity to marine amphipods; T50, chemical concentration corresponding to 50 percent probability of toxicity to marine amphipods; TEC, threshold effect concentration; TEL, threshold effects level; USEPA, U.S. Environmental Protection Agency; WA DOE, Washington Department of Ecology; wt, weight; µg/g-oc, microgram per gram of sediment organic carbon; µg/kg, microgram per kilogram; –, no benchmark available]

Contaminant/benchmark name:	Lower SV[1] NOAA[2] µg/kg dry wt ERL, marine	Upper SV[1] NOAA[2] µg/kg dry wt ERM, marine	Lower SV[1] Canada/Florida[3] µg/kg dry wt TEL, marine	Upper SV[1] Canada/Florida[3] µg/kg dry wt PEL, marine	Lower SV[1] MacDonald[4] µg/kg dry wt TEC	Upper SV[1] MacDonald[4] µg/kg dry wt PEC	Lower SV[1] Field[5] µg/kg dry wt T20	Upper SV[1] Field[5] µg/kg dry wt T50	Upper SV[1] WA DOE[6] µg/kg dry wt AET	EqP[1] USEPA[7] µg/g-oc EqP-acute	EqP[1] USEPA[7] µg/g-oc EqP-chronic
1,2,4-Trichlorobenzene	–	–	–	–	–	–	–	–	>4.8 E	6,100	920
1-Methylnaphthalene	–	–	–	–	–	–	21	94	–	–	–
1-Methylphenanthrene	–	–	–	–	–	–	18	112	–	–	–
2,4,5-Trichlorophenol	–	–	–	–	–	–	–	–	3 I	–	–
2,4,6-Trichlorophenol	–	–	–	–	–	–	–	–	6 I	–	–
2,4-Dichlorophenol	–	–	–	–	–	–	–	–	0.2083	–	–
2,4-Dimethylphenol	–	–	–	–	–	–	–	–	18 N	–	–
2-Chlorophenol	–	–	–	–	–	–	–	–	0.333	–	–
2-Methylnaphthalene	70	670	20.2	201	–	–	21	128	64 E	–	–
4-Bromophenyl phenyl ether	–	–	–	–	–	–	–	–	–	2,300	130
Acenaphthene	16	500	6.71	88.9	–	–	19	116	130 E	–	–
Acenaphthylene	44	640	5.87	128	–	–	14	140	71 E	–	–
Anthracene	85.3	1,100	46.9	245	57.2	845	34	290	280 E	–	–
Benz(a)anthracene	261	1,600	74.8	693	108	1,050	61	466	960 E	–	–
Benzo(a)pyrene	430	1,600	88.8	763	150	1,450	69	520	1,100 E	–	–
Benzo(b)fluoranthene	–	–	–	–	–	–	130	1,107	1,800 EI	–	–
Benzo(ghi)perylene	–	–	–	–	–	–	67	497	670 M	–	–
Benzo(k)fluoranthene	–	–	–	–	–	–	70	537	1,800 EI	–	–
Benzyl n-butyl phthalate	–	–	–	–	–	–	–	–	63 M	15,000	1,100
Biphenyl	–	–	[8]182	[8]2,647	–	–	17	73	–	850	110
Bis(2-ethylhexyl) phthalate	–	–	–	–	–	–	–	–	1,300 I	–	–
Chrysene	384	2,800	108	846	166	1,290	82	650	950 E	–	–
Dibenzo(ah)anthracene	63.4	260	6.22	135	33.0	–	19	113	230 OM	–	–
Dibenzofuran	–	–	–	–	–	–	–	–	110 E	3,700	200
Diethyl phthalate	–	–	–	–	–	–	–	–	6 BL	1,100	63
Dimethyl phthalate	–	–	–	–	–	–	–	–	6 B	–	–
Di-n-butyl phthalate	–	–	–	–	–	–	–	–	58 BL	8,000	1,100
Di-n-octyl phthalate	–	–	–	–	–	–	–	–	61 BL	–	–
Fluoranthene	600	5,100	113	1,494	423	2,230	119	1,034	1,300 E	–	–
Fluorene	19	540	21.2	144	77.4	536	19	114	120 E	–	–

Table 5E. Benchmark values for organic contaminants: supplemental aquatic-life benchmarks for organic contaminants in sediment.—Continued

[Abbreviations: AET, apparent effects threshold; B, bivalve; E, echinoderm larvae; EqP, equilibrium partitioning sediment benchmark; ERL, effects range–low; ERM, effects range–median; I, infaunal community index; L, larvae; M, microtox assay; N, neanthes bioassay; NOAA, National Oceanic and Atmospheric Administration; O, oyster larvae; PEC, probable effect concentration; PEL, probable effect level; SV, screening value; T20, chemical concentration corresponding to 20 percent probability of toxicity to marine amphipods; T50, chemical concentration corresponding to 50 percent probability of toxicity to marine amphipods; TEC, threshold effect concentration; TEL, threshold effects level; USEPA, U.S. Environmental Protection Agency; WA DOE, Washington Department of Ecology; wt, weight; µg/g-oc, microgram per gram of sediment organic carbon; µg/kg, microgram per kilogram; –, no benchmark available]

Benchmark Type:	Lower SV[1]	Upper SV[1]	Lower SV[1]	Upper SV[1]	MacDonald[4]	Upper SV[1]	Lower SV[1]	Upper SV[1]	Upper SV[1]	EqP[1]	EqP[1]
Benchmark source:	NOAA[2]	NOAA[2]	Canada/ Florida[3]	Canada/ Florida[3]	MacDonald[4]	MacDonald[4]	Field[5]	Field[5]	WA DOE[6]	USEPA[7]	USEPA[7]
Units:	µg/kg dry wt	µg/kg dry wt	µg/kg dry wt	µg/kg dry wt	µg/kg dry wt	µg/kg dry wt	µg/kg dry wt	µg/kg dry wt	µg/kg dry wt	µg/g-oc	µg/g-oc
Contaminant/benchmark name:	ERL, marine	ERM, marine	TEL, marine	PEL, marine	TEC	PEC	T20	T50	AET	EqP-acute	EqP-chronic
Hexachlorobenzene	–	–	–	–	–	–	–	–	6 B	–	–
Hexachlorobutadiene	–	–	–	–	–	–	–	–	1.3 E	–	–
Hexachloroethane	–	–	–	–	–	–	–	–	73 BL	1,800	100
Indeno(123-cd)pyrene	–	–	–	–	–	–	68	488	600 M	–	–
m-Cresol plus p-cresol	–	–	–	–	–	–	–	–	[9]100 B	–	–
Naphthalene	160	2,100	34.6	391	176	561	30	217	230 E	–	–
Nitrobenzene	–	–	–	–	–	–	–	–	21 N	–	–
N-Nitrosodiphenylamine	–	–	–	–	–	–	–	–	28 I	–	–
o-Cresol	–	–	–	–	–	–	–	–	8 B	–	–
Pentachlorophenol	–	–	–	–	–	–	–	–	17 B	–	–
Perylene	–	–	–	–	–	–	74	453	–	–	–
Phenanthrene	240	1,500	86.7	544	204	1,170	68	455	660 E	–	–
Phenol	–	–	–	–	–	–	–	–	130 E	–	–
Pyrene	665	2,600	153	1,398	195	1,520	125	932	2,400 E	–	–
Sum high MW PAH	1,700	9,600	655	6,676	–	–	–	–	7,900 E	–	–
Sum low MW PAH	552	3,160	312	1,442	–	–	–	–	1,200 E	–	–
Sum total PAH	4,022	44,792	1,684	16,770	1,610	22,800	–	–	–	–	–

[1] EqP, Equilibrium partitioning sediment benchmark, which is mechanistically based and derived from equilibrium-partitioning theory; Lower SV, empirical screening value below which adverse effects are not expected; Upper SV, empirical screening value above which there is a high probability of adverse effects.

[2] From Long and others (1995).

[3] Except as noted, values are common to both Canada (Canadian Council of Ministers of the Environment, 2011) and Florida Department of Environmental Protection (MacDonald and others, 1996).

[4] From MacDonald and others (2000).

[5] From Field and others (2002).

[6] From Buchman (2008). Value shown is the lowest reliable AET value among available tests, as determined by Buchman (2008). Some values also appear in Barrick and others (1998) or Gries and Waldow (1994). Abbreviations indicate types of bioassays, and are defined in the table headnote.

[7] EqP, Equilibrium partitioning sediment benchmark for either acute or chronic effects on benthic organisms from U.S. Environmental Protection Agency (2004).

[8] Value applies to Florida only (not Canada).

[9] Benchmark applies to para isomer.

Table 6A. Benchmark values for trace and major elements: aquatic-life benchmarks for trace elements in water.

[**Abbreviations**: ANZ, Australian and New Zealand Environment and Conservation Council; BC, British Columbia; CMC, criteria maximum concentration; LOEL, lowest observable effect level; NOAA, National Oceanic and Atmospheric Administration; p, proposed; USEPA, U.S. Environmental Protection Agency; WQC, water-quality criteria; µg/L, microgram0 per liter; –, no benchmark available]

Element	Symbol	Acute, marine (µg/L)[1]	Chronic, marine (µg/L)[1]	Source[2]
Antimony	Sb	[3]1,500	[3]500	NOAA
Arsenic	As	69	36	USEPA WQC
Barium	Ba	1,000	200	BC
Beryllium	Be	1,500	100	BC
Boron[4]	B	–	1,200	BC
Cadmium	Cd	40	8.8	USEPA WQC
Cobalt	Co	–	1	ANZ
Copper[4]	Cu	4.8	3.1	USEPA WQC
Lead	Pb	210	8.1	USEPA WQC
Manganese	Mn	–	100	BC
Mercury	Hg	1.8	0.94	USEPA WQC
Molybdenum	Mo	–	23	ANZ
Nickel	Ni	74	8.2	USEPA WQC; USEPA response
Selenium	Se	290	71	USEPA WQC
Silver	Ag	[5]0.95	–	NOAA
Thallium	Tl	[6]2,130	17	Acute: NOAA; chronic: ANZ
Vanadium	V	–	50	BC; USEPA response
Zinc	Zn	90	81	USEPA WQC

[1]Values are USEPA ambient water-quality criteria supplemented by the lowest of Tier II Species Acute Values or other guidelines, as selected by Buchman (2008). Values were verified (except as noted) in the cited references.

[2]ANZ, Australian and New Zealand Environment and Conservation Council (2000); BC, British Columbia Ministry of Environment (2010); NOAA, Buchman (2008); USEPA WQC, water-quality criteria from U.S. Environmental Protection Agency (2009); USEPA response, USEPA Response to British Petroleum Spill in the Gulf of Mexico from U.S. Environmental Protection Agency (2011a).

[3]p, proposed values from Buchman (2008).

[4]Detected in 1 of 4 field blanks for post-landfall samples, so data were censored prior to comparison with benchmarks (see "Censoring Based on Quality Control Results" in text).

[5]The criterion maximum concentration (CMC, which is USEPA's acute water-quality criterion) was halved to correspond to the 1985 guideline derivation (Buchman, 2008).

[6]USEPA's LOEL; values (unverified) are from Buchman (2008), who compiled LOELs previously published by USEPA.

Table 6B. Benchmark values for trace and major elements: human-health benchmarks (recreational contact) for trace elements in water.

[**Abbreviations**: HH, human health; NC, noncancer effects endpoint; µg/L, micrograms per liter]

Element	Symbol	HH Benchmark (child swimmer)[1] (µg/L)	Cancer/ noncancer
Nickel	Ni	15,000	NC
Vanadium	V	5,400	NC

[1]From U.S. Environmental Protection Agency (2010).

Table 6C. Benchmark values for trace and major elements: aquatic-life benchmarks for trace elements in whole sediment.

[**Abbreviations**: A, amphipod; AET, apparent effects threshold; B, bivalve; E, Echinoderm; ERL, effects range–low; ERM,effects range–median; I, infaunal community index; L, larvae; M, Microtox assay; mg/kg dw, milligram per kilogram dry weight; N, Neanthes bioassay; NOAA, National Oceanic and Atmospheric Administration; O, oyster larvae; PEC, probable effect concentration; PEL, probable effect level; SV, screening value; T20, chemical concentration corresponding to 20 percent probability of toxicity to marine amphipods; T50, chemical concentration corresponding to 50 percent probability of toxicity to marine amphipods; TEC, threshold effect concentration; TEL, threshold effect level; USEPA, U.S. Environmental Protection Agency; WA DOE, Washington Department of Ecology; >, greater than; –, no benchmark]

Benchmark type:			Lower SV[1]	Upper SV[2]	Lower SV[1]	Upper SV[2]	Lower SV[1]	Upper SV[2]	Lower SV[1]	Upper SV[2]	Upper SV[2]
Benchmark source:			NOAA[2]	NOAA[2]	Canada/Florida[3]	Canada/Florida[3]	MacDonald[4]	MacDonald[4]	Field[5]	Field[5]	WA DOE[6]
Element	Symbol	Units/benchmark name:	ERL, marine	ERM, marine	TEL, marine	PEL, marine	TEC	PEC	T20	T50	AET
Aluminum	Al	percent	–	–	–	–	–	–	–	–	1.8 N
Antimony	Sb	mg/kg dw	–	–	–	–	–	–	0.63	2.4	9.3 E
Arsenic	As	mg/kg dw	8.2	70	7.24	41.6	9.79	33	7.4	20	35 B
Barium	Ba	mg/kg dw	–	–	[7]130.1	–	–	–	–	–	48 A
Cadmium	Cd	mg/kg dw	1.2	9.6	0.7	4.2	0.99	4.98	0.38	1.4	3 N
Cobalt	Co	mg/kg dw	–	–	–	–	–	–	–	–	10 N
Chromium	Cr	mg/kg dw	81	370	52.3	160	43.4	111	49	141	62 N
Copper	Cu	mg/kg dw	34	270	18.7	108	31.6	149	32	94	390 MO
Iron	Fe	percent	–	–	–	–	–	–	–	–	22 N
Lead	Pb	mg/kg dw	46.7	218	30.2	112	35.8	128	30	94	400 B
Manganese	Mn	mg/kg dw	–	–	–	–	–	–	–	–	260 N
Mercury	Hg	mg/kg dw	0.15	0.71	0.13	0.70	0.18	1.06	0.14	0.48	0.41 MO
Nickel	Ni	mg/kg dw	20.9	51.6	[8]15.9	[8]42.8	22.7	48.6	15	47	110 EL
Selenium	Se	mg/kg dw	–	–	–	–	–	–	–	–	1 A
Silver	Ag	mg/kg dw	1	3.7	[8]0.73	[8]1.77	–	–	0.23	1.1	3.1 B
Tin	Sn	mg/kg dw	–	–	[9]0.048	–	–	–	–	–	>3.4 N
Vanadium	V	mg/kg dw	–	–	–	–	–	–	–	–	57 N
Zinc	Zn	mg/kg dw	150	410	124	271	121	459	94	245	410 I

[1]Lower SV, empirical screening value below which adverse effects are not expected; Upper SV, empirical screening value above which there is a high probability of adverse effects.

[2]From Long and others (1995).

[3]Except as noted, values are common to both Canada (Canadian Council of Ministers of the Environment, 2011) and Florida Department of Environmental Protection (MacDonald and others, 1996).

[4]From MacDonald and others (2000).

[5]From Field and others (2002).

[6]From Buchman (2008). Value shown is the lowest reliable AET value among available tests, as determined by Buchman (2008). Some values also appear in Barrick and others (1998) or Gries and Waldow (1994). Abbreviations indicate types of bioassays, and are defined in the table headnote.

[7]Based on Screening Level Concentration approach using sensitive species, HC5 (hazardous concentration for 5 percent of species). From Leung and others (2005).

[8]Value applies to Florida only (not Canada).

[9]Based on equilibrium-partitioning approach using the current criterion continuous concentration (CCC, USEPA's chronic ambient water-quality criterion). From Buchman (2008).

Table 6D.   Benchmark values for trace and major elements: national baseline concentrations for trace and major elements in the less than 63-micrometer sediment fraction.

[**Abbreviations**: mg/kg, milligram per kilogram; <, less than; ≤, less than or equal to; ±, plus or minus]

Constituent	Symbol	Units	Baseline minimum[1]	Baseline median[1]	Baseline maximum[1]
Aluminum	Al	percent	4.9	5.9	6.9
Antimony	Sb	mg/kg	0.5	0.7	1.2
Arsenic	As	mg/kg	4.4	6.6	8.8
Barium	Ba	mg/kg	380	490	600
Beryllium	Be	mg/kg	1	1.8	2.6
Cadmium	Cd	mg/kg	0.2	0.37	0.6
Calcium	Ca	percent	0.5	1.8	3.1
Cerium	Ce	mg/kg	54	69	84
Chromium	Cr	mg/kg	45	58	71
Cobalt	Co	mg/kg	8	12	16
Copper	Cu	mg/kg	14	20	26
Iron	Fe	percent	2.2	2.9	3.6
Lanthanum	La	mg/kg	31	39	47
Lead	Pb	mg/kg	14	20	26
Lithium	Li	mg/kg	20	30	40
Magnesium	Mg	percent	0.5	0.9	1.3
Manganese	Mn	mg/kg	480	840	1,200
Mercury	Hg	mg/kg	0.02	0.04	0.06
Molybdenum	Mo	mg/kg	0.7	1	1.3
Nickel	Ni	mg/kg	16	23	30
Phosphorus	P	mg/kg	800	1,000	1,200
Potassium	K	percent	1.2	1.5	1.8
Selenium	Se	mg/kg	0.5	0.65	0.9
Silver	Ag	mg/kg	0.1	0.2	0.3
Sodium	Na	percent	0.3	0.6	0.9
Strontium	Sr	mg/kg	90	150	210
Sulfur	S	percent	0.04	0.08	0.12
Tin	Sn	mg/kg	1.5	2.5	4
Titanium	Ti	percent	0.25	0.33	0.41
Total carbon	TC	percent	1.7	3.3	4.9
Total organic carbon	TOC	percent	1.3	2.4	3.5
Vanadium	V	mg/kg	62	83	104
Zinc	Zn	mg/kg	71	90.5	110

[1]Baseline median, median concentration associated with sites (1) that were predominantly agricultural or undeveloped, (2) where urban land use was ≤5 percent, and (3) where population densities were ≤27 people per square kilometer, calculated from 450 bed-sediment samples collected from streams across the United States. The baseline minimum and baseline maximum values are equivalent to the median baseline ±30 percent median absolute deviation. From Horowitz and Stephens (2008).

National Baseline Concentrations for Trace and Major Elements and Nutrients in Fine Sediment

Trace and major elements and nutrients in the less than 63-μm sediment fraction were compared to national baseline concentrations from Horowitz and Stephens (2008). Although not technically benchmarks, these baseline concentrations can be used to indicate anthropogenic enrichment. Horowitz and Stephens (2008) determined national baseline concentrations for trace and major elements, and some nutrients, in stream sediments collected from agricultural or undeveloped areas or areas with population density less than or equal to 27 people per square kilometer and urban land use less than or equal to 5 percent. These authors found that enrichment of some elements above baseline was associated with urban land use and population density. These elements, in generally decreasing likelihood of enrichment, are lead, mercury, silver, zinc, cadmium, copper, antimony, sulfur, nickel, tin, chromium, arsenic, cobalt, iron, and phosphorus. Horowitz and Stephens (2008) computed the minimum, median, and maximum baseline concentrations for each element, where the range between the minimum and maximum baseline concentrations represents the range of natural geochemical variance. In Horowitz and Stephens (2008), sediment was wet-sieved through a less than 63-μm mesh and subjected to total digestion prior to analysis; thus, these authors determined total concentrations, that is 95 percent or more of the constituent present, in the less than 63-μm sediment fraction. The processing and analytical methods used by these authors are comparable to those used in the present study.

In the present study, the measured concentration of each element was divided by its maximum baseline concentration to obtain a maximum baseline quotient. The maximum baseline concentration is the upper end of the range in baseline values for a given element as determined by Horowitz and Stephens (2008) and listed in table 6D. "Enrichment" above baseline is defined as having the maximum baseline quotient greater than 1, with the following exception. For samples in which the less than 63-μm fraction makes up less than 1 percent of the total sediment, analytical errors are elevated, and there often is insufficient material to run duplicate analyses to determine the degree of precision. In this case, the precision could be as poor as a 100 percent difference, especially at concentrations near the detection level (Arthur J. Horowitz, Research Chemist, USGS, Atlanta, Georgia, written comm., Feb. 3, 2011). For individual samples with less than 1 percent of total sediment in the less than 63-μm fraction, therefore, maximum baseline quotients needed to be elevated above 2 in order to indicate enrichment. By itself, enrichment, as indicated by maximum baseline exceedance, does not necessarily indicate a potential for adverse effects.

Interpretation of Benchmark Exceedances

For organic contaminants, exceedance of either an EqP benchmark or an upper screening value was considered to be an indication of potential toxicity to benthic organisms. Trace elements were considered to be of most concern if they met the following exceedance criteria for both potential toxicity and anthropogenic enrichment: (1) they exceeded one or more upper screening values in whole sediment samples and (2) they were enriched relative to national baseline concentrations in less than 63-μm sediment samples. Because sediment samples were analyzed for total trace-element concentrations, exceedance rates for upper screening values could be overestimated but are not likely to be underestimated; therefore, these rates, and resulting inferences about potential toxicity, can be considered conservatively high.

In addition, for both organic contaminants and trace elements, sediment samples were classified into one of three effect ranges using terminology from Canadian Council of Ministers of the Environment (2001): (1) minimal-effect range, within which adverse biological effects rarely occur (that is, all constituents were below their lower screening values); (2) possible-effect range, within which adverse biological effects occasionally occur (one or more constituents exceeded a lower screening value, but no elements exceeded an upper screening value); or (3) probable-effect range, within which adverse biological effects frequently occur (one or more constituents exceeded an upper screening value).

Data Compilation

Each distinct sampling event is recorded in the USGS database with a unique combination of agency code for the site, station-identification number, sample-collection start date, sample-collection end date, and sample medium. The agency code associated with the samples described in this report is "USGS," and the station-identification numbers are presented in table 1. In the database, sediment samples are assigned sampling-medium designations of either bottom material or soil.

The results for environmental samples from water and sediment can be retrieved from the USGS by supplying the station-identification numbers to one of the following web sites:

- NWISWeb (http://nwis.waterdata.usgs.gov/usa/nwis/qwdata) or
- Water-Quality Data Portal (http://qwwebservices.usgs.gov/portal.html)

The Data Portal provides data in a manner consistent with similar data provided by the STORET database, except that a few of the observational metadata available from NWISWeb are omitted. Samples collected prior to July 15, 2010, are categorized as "pre-landfall," and subsequent samples are categorized as "post-landfall."

Sometimes, one or more constituents in a particular sample were reanalyzed to verify the results or to employ an analytical method with improved sensitivity to low concentrations. When verification reruns were performed, the earliest analytical result is presented in the database, and additional results from a subsequent analysis are preserved in the "result-laboratory" comments field. When a more sensitive method was employed, however, the results from the more sensitive method are presented in the database, and results from the less sensitive method are preserved in the "result-laboratory" comments.

Data Analyses

Data for all analyses described in this report were obtained on March 28, 2011 (March 24 DWH GOM Data Release), and used as received from the participating laboratories without further rounding. Benchmark comparisons were made for all samples, including environmental and field-replicate samples. For most sites, if multiple samples were collected during either the pre-landfall or post-landfall sampling period, one was designated as the primary, or environmental sample, and any others were considered to be replicates for that sampling period. If no primary sample was designated, however, or if the primary sample was missing data for either trace elements or organic analytes, then the replicate sample with the earliest date and time, or with data for the fullest suite of analytes, typically was designated as a primary sample. This "primary-sample" dataset was used for statistical summaries of contaminant occurrence, so that each site was represented only once for each sampling period and analytical method.

A subset of the primary-sample dataset, consisting of paired pre- and post-landfall samples, was used for statistical comparison of pre-landfall to post-landfall sample concentrations at these sites. This "paired-sample" dataset was generated by dropping data for all sites that were sampled during only one sampling period, either pre-landfall or post-landfall. The resulting paired-sample dataset contained exactly two samples per site—one collected during each of the pre-landfall and post-landfall periods.

Detection frequencies and percentile concentrations were determined by using procedures in the statistical software package, SAS 9.2 TS Level 2M3 (SAS Institute Inc., 2009a and 2009b). Summary statistics are presented separately for each chemical class (organic contaminants, or trace and major elements and nutrients) in each sampling medium (water, whole sediment, or less than 63-μm sediment fraction). The detection frequency for a given analyte varies with the sensitivity of the analytical method; for example,

of two methods for a given analyte, the method with the lower reporting level is likely to result in a higher detection frequency. Therefore, to facilitate comparison of detection frequencies between sampling periods and for different contaminants, detection frequencies were calculated at multiple detection thresholds appropriate for the chemical class and sampling medium. These detection thresholds are discussed in detail in the section on "Data Censoring." Briefly, for each analyte, one "optimal" detection threshold was determined to facilitate comparison between pre-landfall and post-landfall samples. In addition, detection frequencies for all analytes within the same contaminant class and sampling medium were computed at each of four common detection thresholds to allow comparison of detection frequencies among analytes. In the context of this study, data censoring refers to the process of distinguishing detections, or quantified values, from nondetections, or censored values; censored datasets are datasets with some portion of the results composed of nondetections.

Percentiles of concentrations were determined in the primary-sample dataset by using one of four methods, depending on the amount of censored data, or nondetections, for a given analyte (fig. 2). For analytes detected in 100 percent of samples, or having no censored data, the SAS UNIVARIATE procedure was used to compute concentration percentiles. For analytes with some, but less than 50-percent, censored data, percentiles were estimated by using the nonparametric Kaplan-Meier method (Helsel, 2005) in the SAS LIFETEST procedure. For analytes with 50- to 80-percent censored data, percentiles were estimated by using a SAS freeware macro, Censored Data Regression on Order Statistics (Helsel, 2005). For analytes with more than 80-percent censored data, all data for that analyte were censored at a common detection threshold, and only the 95th percentile concentration was calculated.

Statistical comparisons of contaminant concentrations in pre-landfall to post-landfall samples were made by using the paired Prentice-Wilcoxon (PPW) test. This test was implemented by using the USGS S-PLUS library version 4.0 (Lorenz and others, 2011) for the statistical software package Spotfire S+ (TIBCO Software, Inc., 2008). The PPW test is appropriate for comparing two groups with matched pairs of data and can be applied to censored datasets. This test evaluates whether there is a difference in the distributions of the two sample groups. To do so, first the data are stacked into one column, a score is computed for each observation (both censored and uncensored data) on the basis of the Kaplan-Meier estimate of the survival function, and then the scores are divided back into the two groups of matched pairs. The PPW test computes the differences between the paired scores and determines whether the sum of these differences is significantly different from zero by using a normal approximation for the test statistic (Helsel, 2005). In this study, the PPW test was performed on the paired-sample dataset, which represents the 48 sites that were sampled during both the pre-landfall and post-landfall periods, as described previously.

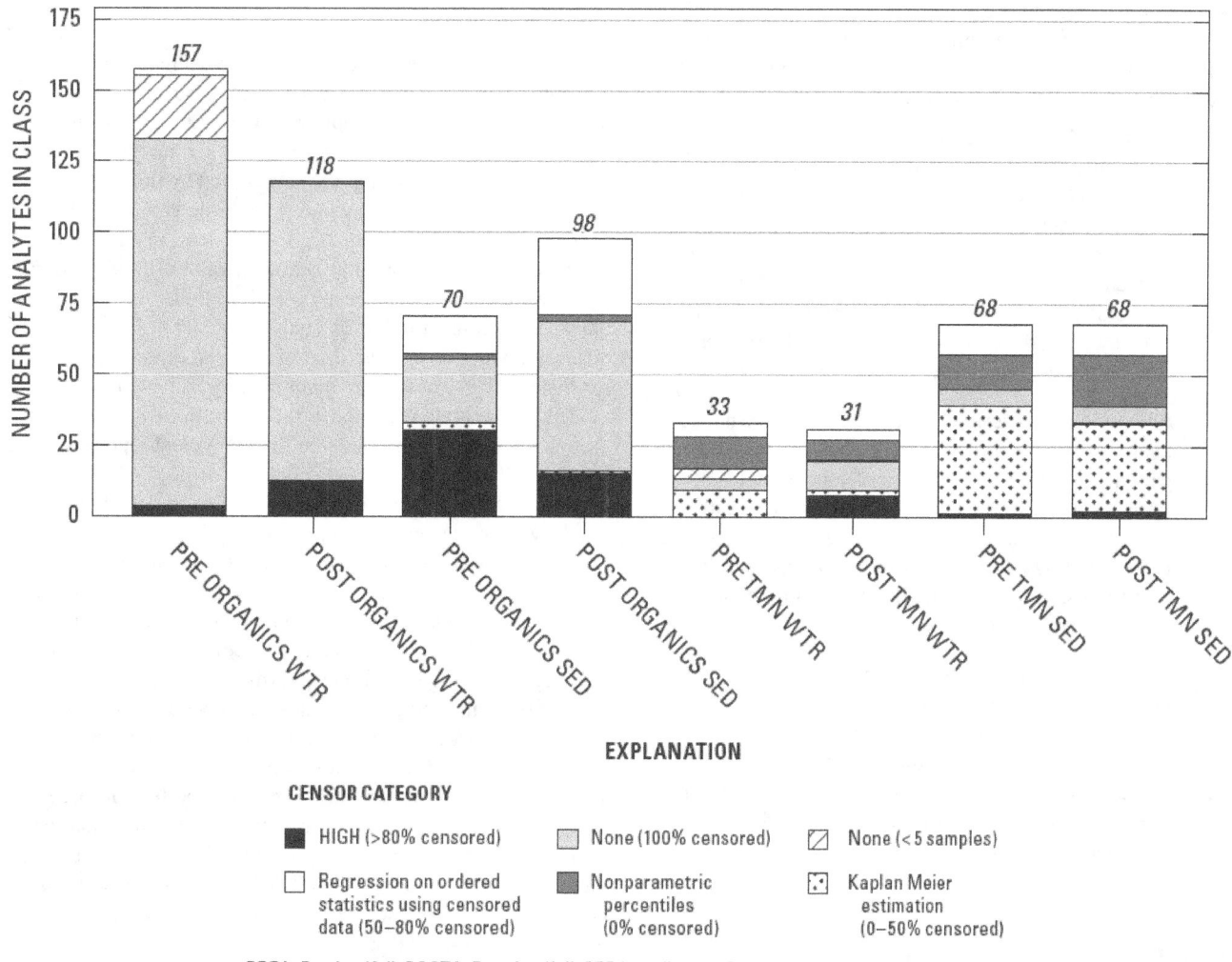

EXPLANATION

CENSOR CATEGORY

■ HIGH (>80% censored)   ▨ None (100% censored)   ▨ None (<5 samples)

□ Regression on ordered   ▨ Nonparametric   ▨ Kaplan Meier
statistics using censored   percentiles   estimation
data (50–80% censored)   (0% censored)   (0–50% censored)

PRE is Pre-landfall; POST is Post-landfall; SED is sediment; TMN is trace and major elements and nutrients;
WTR is water; > is greater than; < is less than; % is percent of data censored

Figure 2. Number of analytes for which percentiles were determined by using four different methods, shown by contaminant class, sampling medium, and sampling period from the 2010 Deepwater Horizon oil spill in the Gulf of Mexico.

Additional PPW tests were used to compare concentrations in pre-landfall and post-landfall samples collected at a subset of 19 paired-sample sites that were identified by Rosenbauer and others (2010) as having geochemical evidence, or a fingerprint, of M-1 well oil in post-landfall samples of sediment, tarballs, or both. At this subset of sites, which is called the "fingerprint-sample" dataset, there is direct evidence from Rosenbauer and others (2010) of residual M-1 well oil at the sites during the post-landfall period.

Benchmark exceedance frequencies were computed by using the Spotfire S+ program. All field samples, including primary and replicate samples, were compared to benchmarks to maximize the information on benchmark exceedance. However, direct comparison between exceedance frequencies for the pre-landfall and post-landfall sampling periods must be qualified because data from the two sampling periods do not represent exactly the same sites. Specifically, 22 pre-landfall sites and 1 post-landfall site were not sampled during the other

sampling period (table 1); also, 20 of the 71 total sites were sampled more than once during one or both sampling periods.

For each combination of contaminant class and sampling medium, Fisher's exact test was used to determine whether the proportion of samples exceeding applicable benchmarks was significantly different ($p < 0.05$) between the pre-landfall and post-landfall sampling periods. This test was performed on the 48 sites in the paired-sample dataset, so that the same sites are represented only once in both sampling periods. In addition, the paired-sample sign test was used to compare selected benchmark exceedance results, such as $\sum TU_i$ and $\sum ESBTU_i$ values, and benchmark exceedances for individual trace elements, between pre-landfall and post-landfall samples in the paired-sample dataset. A nonparametric test with few assumptions, this tests whether the pre-landfall values were generally larger or smaller than the post-landfall values ($p < 0.05$; Helsel and Hirsch, 2002).

For comparison of measured contaminant concentrations to various benchmarks for human health and aquatic life, some data manipulations were necessary because of the nature of the dataset. Specifically, the following apply:

- In this study, trace-element concentrations are reported as total concentrations in water. Because most benchmarks for trace elements are expressed in terms of dissolved concentration in the water column, estimates of these benchmarks as total trace-element concentrations were calculated by using saltwater conversion factors from U.S. Environmental Protection Agency (2011d) prior to comparison with measured concentrations (Buchman, 2008).

- As previously described, water samples were not analyzed for the alkylated PAH groups required for calculation of USEPA benchmarks for mixtures of PAH and BTEX compounds ($\sum TU_i$). Concentrations of each alkylated PAH group were estimated from the corresponding parent PAH concentrations by using multipliers, as specified in the USEPA procedures for benchmark calculation (Mount, 2010).

- For sediment samples, BTEX compounds were not determined, so calculated $\sum ESBTU_i$ values could be slightly low; however, this bias is expected to be minimal because BTEX compounds are volatile, were not detected in weathered M-1 oil (State of Florida Oil Spill Academic Task Force, 2010), and are not expected to persist in sediment (Mount, 2010).

Results and Discussion

The results are presented first for QC analyses because these findings affect interpretation of field sample data. Following the QC data discussion, results are presented separately for each combination of contaminant class and sampling medium. In each case, contaminant occurrence is assessed, statistical comparisons are made between concentrations in pre-landfall and post-landfall samples, and measured concentrations are compared to applicable water- or sediment-quality benchmarks.

Quality-Control Analyses

Analytical results for the various QC samples follow. These results were considered in computing occurrence statistics and making benchmark comparisons, as discussed later in this section.

Blank Samples

Results were available for 166 analytes in at least 4 pre-landfall field blanks. Most of these results were from the USGS NWQL. In addition, results were available from TestAmerica Laboratory in Colorado for six analytes, four of which overlap with analytes determined by the USGS NWQL, and results were available from the USGS OCRL for dissolved organic carbon. Of the 885 total results, 861 (97 percent) were reported as censored values (nondetections). There were 24 quantified results, or detections, reported in blanks, affecting a total of 21 analytes (table 7). Five of the quantified values were less than the highest reporting level for that analyte. These were quantified by using corroborating evidence of analyte presence in the mass spectrogram, even though the concentration was below the typical reporting level for the method.

During the post-landfall sampling period, four field blanks were collected and shipped to the TestAmerica Laboratory in Florida for analysis. Of the 584 total reported results for 146 analytes, 564 (97 percent) were nondetections. There were 20 quantified detections reported for 12 analytes (table 8), of which only 3 are organic contaminants. Ammonia plus organic nitrogen and phosphorus were quantified in each of the four blanks. Trip blanks also were analyzed by the TestAmerica Laboratory during post-landfall sampling. These have limited utility for comparison to environmental samples; however, quantified detections reported for three analytes (table 8) could indicate potential for contamination during laboratory processing and analysis. None of these three analytes were detected in field blanks analyzed at this laboratory. The benzene result was from a blank associated with an environmental sample collected in Louisiana on October 12; the other results were from a blank associated with a sample collected in Florida on October 13.

There was little consistency in blank contamination between sampling periods. Only four analytes—calcium, magnesium, naphthalene, and sodium—were detected in blanks from both the pre-landfall and post-landfall periods. Six analytes detected in the pre-landfall blanks—1,4-dichlorobenzene, 4-chloro-3-methylphenol, arsenic, dichloromethane, ethyl methyl ketone, and silver—were quantified at concentrations less than the reporting level for post-landfall blanks. Similarly, copper was detected in two post-landfall blanks, but at concentrations less than the highest reporting level for pre-landfall blanks. Because of these discrepancies, it was not possible to evaluate differences in incidental contamination between sampling periods. In subsequent data analyses, potential contamination in environmental samples was determined separately for each period by using field blanks collected during that period.

Table 7. Analytes with quantified detections in field blanks collected during the pre-landfall sampling period from the Deepwater Horizon oil spill, Gulf of Mexico, 2010.[1]

[**Abbreviations**: BTEX, benzene, toluene, ethylbenzene, xylenes and related compounds; mg/L, milligram per liter; USGS, U.S. Geological Survey; µg/L, microgram per liter]

Analyte	Units	Number of blanks	Highest reporting level	Number of quantified results	Maximum quantified value	Raised censoring level[2]
USGS National Water-Quality Laboratory						
Organic contaminants						
1,2,3,5-Tetramethylbenzene	µg/L	5	0.08	1	0.032	0.16
1,2,3-Trimethylbenzene	µg/L	5	0.06	1	0.022	0.11
1,2,4-Trimethylbenzene	µg/L	5	0.016	1	0.026	0.13
1,4-Dichlorobenzene	µg/L	6	0.22	1	0.014	0.07
4-Chloro-3-methylphenol	µg/L	6	0.55	1	0.058	0.29
Acetone	µg/L	5	1.7	1	4.5	45
Dichloromethane	µg/L	5	0.019	1	0.64	6.4
Ethyl methyl ketone	µg/L	5	1.6	1	0.49	4.9
Ethylbenzene	µg/L	5	0.018	1	0.031	0.155
Naphthalene	µg/L	6	0.22	1	0.057	0.285
Toluene	µg/L	5	0.009	1	0.083	0.83
Trichloromethane	µg/L	5	0.015	1	1.8	9
Xylene, meta plus para	µg/L	5	0.04	1	0.10	0.5
Xylene, ortho	µg/L	5	0.016	1	0.12	0.6
Trace and major elements						
Arsenic	µg/L	[3]4	0.09	1	0.15	0.75
Calcium	mg/L	[4]4	0.02	1	0.02	0.1
Lithium	µg/L	5	0.04	1	0.23	1.15
Magnesium	mg/L	5	0.012	1	0.013	0.065
Silver	µg/L	5	0.12	1	0.57	2.85
Sodium	mg/L	5	0.36	1	0.41	2.05
USGS Organic Carbon Research Laboratory						
Dissolved organic carbon	mg/L	5	0.7	4	0.70	3.5

[1]Blanks for BTEX compounds, gasoline range organics, and diesel range organics were submitted to the TestAmerica Laboratory in Colorado; all results were censored.

[2]The censoring level was raised to 5 times the maximum quantified value or, for common laboratory contaminants, to 10 times the maximum quantified value.

[3]One result with an elevated reporting level of 1.35 µg/L was excluded.

[4]One result with an elevated reporting level of 0.06 µg/L was excluded.

Sediment-equipment rinsate blank results were available for 146 analytes from the TestAmerica Laboratory in Florida and for one analyte from the USGS OCRL. Of the 389 total reported results, 365 (94 percent) are nondetections. There were 24 quantified detections reported for 14 analytes (table 9). Similar to results for the post-landfall field blanks, ammonia plus organic nitrogen and phosphorus were detected in each of the sediment-equipment rinsate blanks. Naphthalene and toluene were the only organic compounds detected. The extremely high concentration of boron in one blank could have been caused by residue from a cleaning solution that was used on the sampling equipment. If so, the potential for contamination of a sediment sample collected by using this equipment is probably much less than the concentration in a blank-water rinse. Any residue would likely be washed away during field rinsing of the equipment.

Table 8.   Analytes with quantified detections in field and trip blanks collected during the post-landfall sampling period and analyzed at the TestAmerica Laboratory in Pensacola, Florida, from the 2010 Deepwater Horizon oil spill in the Gulf of Mexico.

[**Abbreviations**: mg/L, milligrams per liter; µg/L, micrograms per liter; –, no results were censored]

Analyte	Units	Number of blanks	Highest reporting level	Number of quantified results	Maximum quantified value	Raised censoring level[1]
Field blanks						
Organic contaminants						
Diesel range organics	µg/L	4	46	1	50	250
Diethyl phthalate	µg/L	4	0.26	1	0.42	4.2
Naphthalene	µg/L	4	0.15	1	0.16	0.8
Trace and major elements, and nutrients						
Ammonia plus organic nitrogen as N	mg/L	4	0.02	4	1.7	8.5
Boron	µg/L	4	5	1	10	50
Calcium	mg/L	4	0.03	1	0.072	0.36
Copper	µg/L	4	2.0	2	2.2	11
Magnesium	mg/L	4	0.03	1	0.2	1
Mercury	µg/L	4	0.07	2	0.18	0.9
Phosphorus as P	mg/L	4	–	4	0.13	0.65
Potassium	mg/L	4	0.1	1	0.16	0.8
Sodium	mg/L	4	0.5	1	2.2	11
Trip blanks						
Organic contaminants						
Benzene	µg/L	31	0.34	1	0.42	2.1
Dichloromethane	µg/L	31	1	1	3.1	15.5
Trichlorofluoromethane	µg/L	31	0.52	1	0.62	3.1

[1]The censoring level was raised to 5 times the maximum quantified value in blanks or, for common laboratory contaminants, to 10 times the maximum quantified value in blanks.

Table 9.   Analytes with quantified detections in sediment-equipment rinsate blanks from the Deepwater Horizon oil spill, Gulf of Mexico, 2010.

[**Abbreviations**: mg/L, milligrams per liter; USGS, U.S. Geological Survey; µg/L, micrograms per liter; –, not applicable]

Analyte	Units	Number of blanks	Highest reporting level	Number of quantified results	Maximum quantified value
TestAmerica Laboratory—Florida					
Organic contaminants					
Naphthalene	µg/L	3	0.15	1	0.76
Toluene	µg/L	2	0.70	1	8.7
Trace and major elements, and nutrients					
Ammonia plus organic nitrogen as N	mg/L	3	–	3	1.6
Boron	µg/L	3	5.0	2	500
Calcium	mg/L	3	0.030	2	0.088
Copper	µg/L	3	2.0	1	4.4
Magnesium	mg/L	3	0.030	1	0.22
Manganese	µg/L	3	1.0	1	1.5
Mercury	µg/L	3	0.070	1	0.11
Phosphorus as P	mg/L	3	–	3	0.18
Potassium	µg/L	3	0.10	1	0.25
Sodium	mg/L	3	0.50	2	2.7
Zinc	µg/L	3	8.0	1	19
USGS Organic Carbon Research Laboratory					
Dissolved organic carbon	mg/L	4	–	4	0.50

Field Replicates

Replicate samples were analyzed at all laboratories used in this study (table 4). Replicate-sample data analysis requires quantified detections for at least two samples in a set in order to compute a standard error. In this study, many analytes, particularly organic compounds in water, were not detected in most or all replicate samples. Only those analytes with at least two quantified detections in at least four replicate sets were included in this analysis of variability.

Generally, the number of replicate sets was too small to evaluate variability over low and high ranges of concentration, so variability was simply estimated as the average RSD. This can be considered a conservatively high estimate of variability because RSD values for low-concentration replicates typically are much higher than the average for high-concentration replicates. In subsequent interpretation of environmental data, variability was noted as

a possible source of uncertainty for any contaminant with a replicate RSD greater than 10 percent for water or 20 percent for sediment.

Replicate water samples collected during the pre-landfall period were analyzed at the USGS NWQL, the USGS OCRL, and the TestAmerica Laboratory in Colorado. The USGS OCRL also analyzed replicate water samples from the post-landfall period, and these were combined with the pre-landfall samples for data analysis. Replicate sets with quantified detections were available for only 21 analytes: 2 organic compounds and 17 major ions, nutrients, or trace elements from the USGS NWQL, plus dissolved organic carbon and dissolved nitrogen from the USGS OCRL. The number of pre-landfall replicate sets ranged from 4 to 27, depending on the analyte, and the resulting mean RSD ranged from about 1 percent to almost 19 percent (table 10). The mean RSD exceeded 10 percent for 8 of the 21 analytes in pre-landfall water samples.

Table 10. Mean relative standard deviation for water analytes with quantified detections in at least two samples in at least four sets of replicate water samples from the Deepwater Horizon oil spill, Gulf of Mexico, 2010.[1]

[**Abbreviations**: mg/L, milligrams per liter; RSD, relative standard deviation; USGS, U.S. Geological Survey; µg/L, micrograms per liter; –, not applicable]

Analyte	Units	USGS National Water Quality Laboratory (pre-landfall)		TestAmerica Laboratory, Florida (post-landfall)	
		Number of replicate sets	Mean RSD (percent)	Number of replicate sets	Mean RSD (percent)
Organic contaminants					
Isophorone	µg/L	6	7.59	–	–
Dissolved organic carbon[2]	mg/L	13	3.81	–	–
Trace and major elements, and nutrients					
Aluminum	µg/L	10	13.87	5	9.23
Ammonia plus organic nitrogen as N	mg/L	27	12.76	7	16.16
Ammonia as N	mg/L	22	12.66	–	–
Arsenic	µg/L	26	8.68	–	–
Barium	µg/L	26	4.59	7	8.13
Boron	µg/L	–	–	7	0.83
Calcium	mg/L	26	2.63	7	1.48
Cobalt	µg/L	8	11.15	–	–
Iron	µg/L	21	18.72	4	19.18
Lithium	µg/L	26	2.98	–	–
Magnesium	mg/L	26	2.78	7	0.56
Manganese	µg/L	21	14.06	5	12.09
Mercury	µg/L	–	–	4	29.63
Molybdenum	µg/L	26	1.57	–	–
Phosphorus as P	mg/L	18	15.38	7	9.05
Potassium	mg/L	26	3.06	7	4.96
Selenium	µg/L	4	5.77	–	–
Sodium	mg/L	26	1.01	7	2.10
Strontium	µg/L	26	2.78	–	–
Dissolved nitrogen[2]	mg/L	13	5.64	–	–

[1]Replicates collected during the pre-landfall period were submitted to the TestAmerica Laboratory in Colorado for analysis of diesel range organics, but only one set had more than one quantified result.

[2]Analyzed by USGS Organic Carbon Research Laboratory; samples collected during both sampling periods.

Replicate water samples collected during the post-landfall period were analyzed at the TestAmerica Laboratory in Florida. Quantified detections were available to assess the variability for 12 major ions, nutrients, or trace elements. The number of replicates sets ranged from 4 to 7, depending on the analyte, and the mean RSD ranged from less than 1 percent to almost 30 percent (table 10). The mean RSD exceeded 10 percent for 4 of the 12 analytes in post-landfall water samples.

Replicate sediment samples collected during the pre-landfall period were analyzed at the USGS NWQL, the USGS SCL, and the TestAmerica Laboratories in Colorado and Vermont (table 4). Samples collected during the post-landfall period were analyzed at the USGS SCL and the TestAmerica Laboratories in Florida and Vermont. There were too few detections in replicate data from the USGS NWQL and the TestAmerica Laboratory in Florida to compute representative mean RSD values. For the other laboratories, replicate data from both sampling periods were combined for this analysis. Quantified results were available for 15 organic contaminants from the TestAmerica laboratories and for 31 trace and major elements and nutrients from the USGS SCL. Analyses at the SCL included both whole sediment and the less than 63-µm sediment fraction. Mean RSD values were computed for all 31 analytes in the less than 63-µm sediment fraction, but quantified detections were available to compute mean RSD values for only two analytes—molybdenum and tin—in the whole-sediment samples.

Table 11 lists mean RSD values for organic contaminants in whole sediment and for trace and major elements and nutrients in the less than 63-µm sediment fraction. For organic contaminants, the number of replicate sets ranged from 5 to 17, depending on the contaminant, and the resulting mean RSD ranged from about 9 percent to more than 47 percent. Mean RSD exceeded 20 percent for 12 of the 15 organic contaminants in sediment. For trace and major elements and nutrients, the number of replicate sets ranged from 4 to 17, and the mean RSD ranged from about 2 percent to more than 28 percent. Mean RSD exceeded 20 percent for 4 of the 31 constituents in sediment.

Matrix Spikes

The USGS NWQL spiked 85 organic compounds in 5 separate water-matrix samples collected during the pre-landfall period. Mean recovery for individual analytes ranged from about 52 to 134 percent. The lowest recovery was for dichlorodifluoromethane; otherwise, all recoveries were greater than 60 percent. The highest recovery was for acetone, which is a common laboratory contaminant and was measured at 4.5 micrograms per liter (µg/L) in one field blank analyzed at the USGS NWQL. The next highest recovery was only about 110 percent. Thus, almost all recoveries for this group of spikes were between 60 and 110 percent.

Table 11.   Mean relative standard deviation for analytes with quantified detections in at least two samples in at least four sets of replicate sediment samples from the Deepwater Horizon oil spill, Gulf of Mexico, 2010.[1,2]

[**Abbreviations**: mg/kg, milligrams per liter; RSD, relative standard deviation; USGS, U.S. Geological Survey; µg/kg, micrograms per kilogram]

Analyte	Units	Number of replicate sets	Mean RSD (percent)
Selenium	mg/kg	16	12.05
Sodium	percent	17	28.44
Strontium	mg/kg	17	10.28
Sulfur	percent	17	17.17
Tin	mg/kg	5	23.84
Titanium	percent	15	7.94
Vanadium	mg/kg	16	8.30
Zinc	mg/kg	17	16.97

[1]Replicates collected during the pre-landfall period were submitted to the USGS National Water Quality Laboratory for analysis of several organic compounds, but none had more than one quantified result in more than three sets.

[2]Replicates collected during the post-landfall period were submitted to the TestAmerica Laboratory in Florida for analysis of oil and grease, but only 3 sets had more than one quantified result.

The TestAmerica Laboratory in Florida prepared duplicate spikes for 107 organic compounds and 24 trace elements in 5 water-matrix samples during the post-landfall period. Mean recovery for individual analytes ranged from about 19 to 124 percent. The lowest mean recoveries were for 3, 3'-dichlorobenzidine at 19.2 percent and N-nitrosodiphenylamine at 43.4 percent; otherwise, all recoveries were greater than 52 percent. The highest recoveries were for aluminum at 124 percent and mercury at 117 percent. Mercury also was found in two field blanks at the TestAmerica Laboratory in Florida, at a maximum concentration of 0.18 µg/L; therefore, the high recovery could have been due to contamination.

Spikes in water-matrix samples at the two laboratories had 41 analytes in common. Differences in recoveries were generally small—less than 17 percent for all but five analytes.

The USGS NWQL spiked 37 organic compounds into 4 separate sediment-matrix samples collected during the pre-landfall period. Mean recovery for individual analytes ranged from about 23 to 62 percent. The lowest mean recoveries were for 1,2,4-trichlorobenzene at 22.9 percent and naphthalene at 33.7 percent; otherwise, all recoveries were greater than 44 percent. The TestAmerica Laboratory in Florida prepared duplicate spikes for 59 organic compounds in either 3 or 4 sediment-matrix samples during the post-landfall period. Mean recovery for individual analytes ranged from 43 to about 88 percent. The lowest mean recoveries were for N-nitrosodiphenylamine at 43.0 percent and 4-chloroaniline at 56.2 percent; otherwise, all recoveries were greater than 61 percent. Spikes at the 2 laboratories had 18 analytes in common; mean recoveries in spikes from the TestAmerica Laboratory in Florida were consistently higher by about 13 to 35 percent.

Analytes with low spike recovery could also have a low bias in environmental-sample results. In the present study, recovery is considered to be within acceptable limits if it is between 70 and 115 percent for organic analytes in water samples and between 50 and 115 percent for organic analytes in sediment samples. Table 12 provides a list of analytes with less than 70 percent or more than 115 percent recovery

in water spikes, or with less than 50 percent or more than 115 percent recovery in sediment spikes. Concentrations reported for these analytes in environmental samples could be substantially lower than their true concentrations. Analytes with overly high spike recovery could have a high bias in environmental-sample results, possibly due to laboratory contamination. This condition primarily affects acetone in water samples analyzed at the USGS NWQL and aluminum and mercury in water samples analyzed by the TestAmerica Laboratory in Florida. Concentrations were not recovery-corrected, but analytes with exceptionally low or high recovery are footnoted in tables within this report.

Table 12. Analytes with less than 70 percent or more than 115 percent recovery in water matrix spikes, or with less than 50 percent or more than 115 percent recovery in sediment matrix spikes from the Deepwater Horizon oil spill, Gulf of Mexico, 2010.[1]

[**Abbreviations**: mg/L, milligram per liter; P, phosphorus; USGS, U.S. Geological Survey; μg/L, microgram per liter; μg/kg, microgram per kilogram; –, no spiked samples or mean recovery greater than 70 percent]

Analyte	Units	USGS National Water Quality Laboratory (pre-landfall)		TestAmerica Laboratory, Florida (post-landfall)	
		Number of spikes	Mean recovery (percent)	Number of spike sets	Mean recovery (percent)
Water					
1,1,2-Trichloro-1,2,2-trifluoroethane	μg/L	5	67.9	3	93.0
2,4-Dimethylphenol	μg/L	–	–	5	59.3
3,3'-Dichlorobenzidine	μg/L	–	–	5	19.2
4-Isopropyltoluene	μg/L	5	69.3	–	–
4-Nitroaniline	μg/L	–	–	5	61.3
4-Nitrophenol	μg/L	–	–	5	68.9
Acetone	μg/L	5	134	–	–
Aluminum	μg/L	–	–	3	124
Benzo[a]pyrene	μg/L	–	–	5	67.3
Carbon disulfide	μg/L	5	63.4	3	88.2
Dichlorodifluoromethane	μg/L	5	52.3	3	97.0
Hexachlorobutadiene	μg/L	5	60.7	5	74.8
Hexachlorocyclopentadiene	μg/L	–	–	5	53.3
Hexachloroethane	μg/L	5	76.7	5	66.2
Mercury	μg/L			3	117
n-Butylbenzene	μg/L	5	65.6	–	–
N-Nitrosodiphenylamine	μg/L	–	–	5	43.4
n-Propylbenzene	μg/L	5	67.9	–	–
Styrene	μg/L	5	61.2	3	88.8
Phosphorus as P	mg/L	–	–	3	52.5
Sediment					
1,2,4-Trichlorobenzene	μg/kg	4	22.9	–	–
1,6-Dimethylnaphthalene	μg/kg	4	47.9	–	–
2,6-Dimethylnaphthalene	μg/kg	4	47.4	–	–
2-Ethylnaphthalene	μg/kg	4	46.3	–	–
Acenaphthylene	μg/kg	4	47.0	3	75.7
Naphthalene	μg/kg	4	33.7	3	69.0
N-Nitrosodiphenylamine	μg/kg	–	–	3	43.0
Pentachloronitrobenzene	μg/kg	4	44.6	–	–

[1]Environmental samples also were analyzed at the USGS Sediment Chemistry Laboratory and the TestAmerica Laboratories in Colorado and Vermont, but no matrix-spike results were reported.

Data Censoring

If an analyte cannot be reliably quantified—for example, if the measured value is less than the detection level or if there is no evidence, such as from mass spectra, that the analyte is present—then the analytical result reported by the laboratory is censored, that is, reported as less than a specified concentration, called a reporting level. In statistical terms, this practice results in censored data, which require special methods for data analysis. Many constituents were

not quantified in any environmental sample collected for this study. Table 13 lists 114 organic contaminants that were censored—that is, not detected—in every water sample, and table 14 lists 51 organic contaminants and 3 trace elements that were censored in every sediment sample. These constituents were excluded from subsequent statistical tests and benchmark comparisons. In addition, concentrations of some detected analytes in environmental samples were subject to post-laboratory censoring on the basis of the QC analysis results, as described in the next subsection.

Table 13. Constituents that were not detected in any environmental water samples analyzed from the Deepwater Horizon oil spill, Gulf of Mexico, 2010.

[**Abbreviations**: BTEX, benzene, toluene, ethylbenzene, xylene and related compounds; CARB, carbon, including petroleum hydrocarbons and hexane-extractable oil and grease; PAH, polycyclic aromatic hydrocarbon, including parent and alkylated compounds; SVOC, semi-volatile organic compound (excluding PAHs); TME, trace and major elements; VOC, volatile organic compound; – not analyzed for that time period]

Analytes	Chemical class	Number of samples less than the reporting level	
		Pre-landfall	Post-landfall
Organic constituents			
1,1,1,2-Tetrachloroethane	VOC	60	–
1,1,1-Trichloroethane	VOC	62	48
1,1,2,2-Tetrachloroethane	VOC	62	48
1,1,2-Trichloro-1,2,2-trifluoroethane	VOC	62	48
1,1,2-Trichloroethane	VOC	62	48
1,1-Dichloroethane	VOC	62	48
1,1-Dichloroethene	VOC	62	48
1,1-Dichloropropene	VOC	60	–
1,2,3,4-Tetramethylbenzene	VOC	60	–
1,2,3,5-Tetramethylbenzene	VOC	60	–
1,2,3-Trichlorobenzene	VOC	60	–
1,2,3-Trichloropropane	VOC	60	–
1,2,3-Trimethylbenzene	VOC	60	–
1,2,4-Trichlorobenzene	VOC	68	48
1,2,4-Trimethylbenzene	VOC	60	–
1,2-Dibromo-3-chloropropane	VOC	62	48
1,2-Dibromoethane	VOC	62	48
1,2-Dichlorobenzene	VOC	68	48
1,2-Dichloroethane	VOC	62	48
1,2-Dichloropropane	VOC	62	48
1,2-Diphenylhydrazine	SVOC	65	–
1,3-Dichlorobenzene	VOC	68	48
1,3-Dichloropropane	VOC	60	–
1,4-Dichlorobenzene	VOC	68	48
2,2-Dichloropropane	VOC	60	–
2,4,5-Trichlorophenol	SVOC	2	48
2,4,6-Trichlorophenol	SVOC	67	48
2,4-Dichlorophenol	SVOC	67	48
2,4-Dimethylphenol	SVOC	67	48
2,4-Dinitrophenol	SVOC	64	48
2,4-Dinitrotoluene	VOC	67	48
2-Chloronaphthalene	PAH	67	48
2-Chlorophenol	SVOC	67	48
2-Chlorotoluene	VOC	60	–
2-Methyl-4,6-dinitrophenol	SVOC	67	48

Table 13. Constituents that were not detected in any environmental water samples analyzed from the Deepwater Horizon oil spill, Gulf of Mexico, 2010.—Continued

[**Abbreviations**: BTEX, benzene, toluene, ethylbenzene, xylene and related compounds; CARB, carbon, including petroleum hydrocarbons and hexane-extractable oil and grease; PAH, polycyclic aromatic hydrocarbon, including parent and alkylated compounds; SVOC, semi-volatile organic compound (excluding PAHs); TME, trace and major elements; VOC, volatile organic compound; – not analyzed for that time period]

Analytes	Chemical class	Number of samples less than the reporting level	
		Pre-landfall	Post-landfall
Organic constituents—Continued			
2-Methylnaphthalene	PAH	2	48
2-Naphthylamine	SVOC	2	48
2-Nitrophenol	SVOC	67	48
3,3'-Dichlorobenzidine	SVOC	67	48
3-Chloropropene	VOC	60	–
3-Nitroaniline	SVOC	2	48
4-Bromophenyl phenyl ether	SVOC	67	48
4-Chloro-3-methylphenol	SVOC	67	48
4-Chloroaniline	SVOC	2	48
4-Chlorophenyl phenyl ether	SVOC	67	48
4-Chlorotoluene	VOC	60	–
4-Isopropyltoluene	VOC	60	–
4-Nitroaniline	SVOC	2	48
4-Nitrophenol	SVOC	67	48
Acetone	VOC	62	48
Acetophenone	SVOC	2	48
Acrylonitrile	VOC	60	–
Atrazine	SVOC	2	48
Benzaldehyde	SVOC	2	48
Benzyl *n*-butylphthalate	PAH	67	48
Biphenyl	SVOC	2	48
Bis(2-Chloroisopropyl) ether	SVOC	67	48
Bis-2-Chloroethoxymethane	VOC	67	48
Bromobenzene	VOC	60	–
Bromochloromethane	VOC	60	–
Bromodichloromethane	VOC	62	48
Bromoethene	VOC	60	–
Bromomethane	VOC	62	48
Caprolactam	SVOC	2	48
Carbazole	SVOC	2	48
Chlorobenzene	VOC	62	48
Chloroethane	VOC	62	48
cis-1,2-Dichloroethene	VOC	62	48
cis-1,3-Dichloropropene	VOC	62	48
Cyclohexane	VOC/BTEX	2	48
Dibenzofuran	SVOC	2	48
Dibromochloromethane	VOC	62	48
Dichlorodifluoromethane	VOC	62	48
Dichloromethane	VOC	62	48
Diethyl ether	VOC	60	–
Diisopropyl ether	VOC	60	–
Dimethyl phthalate	VOC	67	48
Ethyl methacrylate	VOC	60	–
Ethyl methyl ketone	VOC	62	48
Ethylbenzene	VOC/BTEX	63	48
Gasoline range organics	CARB	1	–
Hexachlorobenzene	SVOC	67	48

Table 13.   Constituents that were not detected in any environmental water samples analyzed from the Deepwater Horizon oil spill, Gulf of Mexico, 2010.—Continued

[**Abbreviations**: BTEX, benzene, toluene, ethylbenzene, xylene and related compounds; CARB, carbon, including petroleum hydrocarbons and hexane-extractable oil and grease; PAH, polycyclic aromatic hydrocarbon, including parent and alkylated compounds; SVOC, semi-volatile organic compound (excluding PAHs); TME, trace and major elements; VOC, volatile organic compound; – not analyzed for that time period]

Analytes	Chemical class	Number of samples less than the reporting level	
		Pre-landfall	Post-landfall
Organic constituents—Continued			
Hexachlorobutadiene	VOC	68	48
Hexachlorocyclopentadiene	SVOC	67	48
Hexachloroethane	VOC	68	48
Iodomethane	VOC	60	–
Isobutyl methyl ketone	VOC	62	48
Isopropylbenzene	VOC/BTEX	62	48
m-plus *p*-Cresol	SVOC	2	48
m-plus *p*-Xylene	VOC	60	–
Methyl acetate	VOC	2	48
Methyl acrylate	VOC	60	–
Methyl acrylonitrile	VOC	60	–
Methyl methacrylate	VOC	60	–
Methyl tert-butyl ether	VOC	62	48
Methyl tert-pentyl ether	VOC	60	–
Methylcyclohexane	VOC	2	48
Naphthalene	PAH	68	48
n-Butyl methyl ketone	VOC	62	48
n-Butylbenzene	PAH	60	–
Nitrobenzene	SVOC	67	48
N-Nitrosodimethylamine	SVOC	65	–
N-Nitrosodi-*n*-propylamine	SVOC	67	48
N-Nitrosodiphenylamine	SVOC	67	48
o-Cresol	SVOC	2	48
Oil and grease	CARB	–	48
o-Xylene	VOC	60	–
sec-Butylbenzene	VOC	60	–
Styrene	VOC	62	48
tert-Butyl ethyl ether	VOC	60	–
tert-Butylbenzene	VOC	60	–
Tetrachloroethene	VOC	62	48
Tetrachloromethane	VOC	62	48
Tetrahydrofuran	VOC	60	–
trans-1,2-dichloroethene	VOC	62	48
trans-1,3-dichloropropene	VOC	62	48
trans-1,4-Dichloro-2-butene	VOC	60	–
Trichloroethene	VOC	62	48
Trichlorofluoromethane	VOC	62	48
Vinyl chloride	VOC	62	48
Trace and major elements			
Antimony	TME	2	48
Mercury	TME	–	48
Silver	TME	63	48
Thallium	TME	2	48

Table 14. Constituents that were not detected in any environmental sediment samples analyzed from the Deepwater Horizon oil spill, Gulf of Mexico, 2010.

[Sediment samples are whole sediment unless specified otherwise. **Abbreviations**: CARB, carbon, including petroleum hydrocarbons and hexane-extractable oil and grease; PAH, polycyclic aromatic hydrocarbon; SVOC, semi-volatile organic compound (excluding PAHs); TME, trace and major elements; VOC, volatile organic compound; <, less than; – not analyzed for that time period]

Analytes	Chemical class	Number of samples less than the reporting level	
		Pre-landfall	Post-landfall
Organic constituents			
1,2,4-Trichlorobenzene	SVOC	68	–
1-Methylfluorene	PAH	69	–
2,4,5-Trichlorophenol	SVOC	–	48
2,4,6-Trichlorophenol	SVOC	–	48
2,4-Dichlorophenol	SVOC	–	48
2,4-Dimethylphenol	SVOC	–	48
2,4-Dinitrophenol	SVOC	–	48
2,4-Dinitrotoluene	SVOC	–	48
2,6-Dinitrotoluene	SVOC	–	48
2-Chloronaphthalene	PAH	–	48
2-Chlorophenol	SVOC	–	48
2-Methyl-4,6-dinitrophenol	SVOC	–	48
2-Methylanthracene	PAH	69	–
2-Naphthylamine	PAH	–	48
2-Nitrophenol	SVOC	–	48
3,3'-Dichlorobenzidine	SVOC	–	48
3-Nitroaniline	SVOC	–	48
4-Bromophenyl phenyl ether	SVOC	–	48
4-Chloro-3-methylphenol	SVOC	–	48
4-Chloroaniline	SVOC	–	48
4-Chlorophenyl phenyl ether	SVOC	–	48
4-Nitroaniline	SVOC	–	48
4-Nitrophenol	SVOC	–	48
Acetophenone	SVOC	–	48
Atrazine	SVOC	–	48
Benzaldehyde	SVOC	–	48
Benzyl *n*-butylphthalate	SVOC	–	48
Bis(2-Chloro-1-methylethyl) ether	SVOC	–	48
Bis(2-Chloroethoxy)methane	SVOC	–	48
Bis(2-Chloroethyl) ether	SVOC	–	48
Caprolactam	SVOC	–	48
Dibenzofuran	SVOC	–	48
Diesel range organics (C10-C36)	CARB	2	–
Diethylphthalate	SVOC	69	48
Dimethylphthalate	SVOC	–	48
Di-*n*-butyl phthalate	SVOC	–	48
Di-*n*-octyl phthalate	SVOC	–	48
Hexachlorobenzene	SVOC	69	48
Hexachlorobutadiene	VOC	–	48
Hexachlorocyclopentadiene	SVOC	–	48
Hexachloroethane	SVOC	–	48
Isophorone	SVOC	–	48
m-plus *p*-Cresol	SVOC	–	48
Nitrobenzene	SVOC	–	48

Table 14.   Constituents that were not detected in any environmental sediment samples analyzed from the Deepwater Horizon oil spill, Gulf of Mexico, 2010.—Continued

[Sediment samples are whole sediment unless specified otherwise. **Abbreviations**: CARB, carbon, including petroleum hydrocarbons and hexane-extractable oil and grease; PAH, polycyclic aromatic hydrocarbon; SVOC, semi-volatile organic compound (excluding PAHs); TME, trace and major elements; VOC, volatile organic compound; <, less than; – not analyzed for that time period]

Analytes	Chemical class	Number of samples less than the reporting level	
		Pre-landfall	Post-landfall
Organic constituents—Continued			
N-Nitrosodi-*n*-propylamine	SVOC	–	48
N-Nitrosodiphenylamine	SVOC	–	48
o-Cresol	SVOC	–	48
Pentachloroanisole	SVOC	69	–
Pentachloronitrobenzene	SVOC	69	–
Pentachlorophenol	SVOC	–	48
Phenanthridine	SVOC	69	–
Phenol	SVOC	–	48
Trace and major elements			
Thallium, in <63-micrometer sediment	TME	63	37
Thallium	TME	70	49
Uranium	TME	70	49

Censoring on the Basis of Quality-Control Results

For analytes detected in laboratory, field, or trip blanks, concentrations in environmental samples were censored at raised censoring levels on the basis of guidance from the U.S. Environmental Protection Agency (1989, pages 16–17 in chapter 5). Field and trip blanks were available for water samples only, and laboratory reagent blanks were available for both water and sediment. For analytes detected in these blanks, a raised censoring level equal to five times the maximum concentration detected in the blanks was applied to results in any associated environmental samples. This raised censoring level ensures that a reported detection has a high probability of reflecting the actual concentration in the environmental sample, rather than the effect of incidental contamination from sampling and analysis procedures. Quantified results less than this raised censoring level were changed to censored values and reported as less than the quantified value. For example, naphthalene was detected in a post-landfall field blank, so was censored at a raised censoring level of 0.8. A quantified result of 0.5 would be censored to less than 0.5, indicating that the environmental contaminant concentration in that sample is no more than 0.5, but it could be less. For a few common laboratory contaminants—acetone, dichloromethane, diethyl phthalate, methyl ethyl ketone, and toluene—the censoring level was raised to 10 times the maximum concentration detected in the blank.

Four organic contaminants in sediment, four trace or major elements in water, and two nutrients in water had one or more detections in laboratory reagent blanks. Concentrations in all environmental samples, however, were more than five times the reagent blank concentration, except for the two nutrients in water—ammonia plus organic

nitrogen and phosphorus. Because a reagent blank sample is associated with a particular set of environmental samples, censoring for reagent-blank contamination was applied only to those environmental samples that had contamination in the associated reagent blank. Therefore, results for ammonia plus organic nitrogen were censored in 8 of the 48 post-landfall water samples, and phosphorus was censored in 26 of the 48 post-landfall water samples and in 15 of the 68 pre-landfall water samples.

In this study, it was not possible to associate a particular field blank with each environmental sample, so an alternative procedure had to be used to estimate potential contamination. One option was to determine the statistical distribution of concentrations in a set of representative blanks and assume this same distribution applied to potential contamination in the environmental samples (Mueller and Titus, 2005; Apodaca and others, 2006). This procedure requires more than 20 blanks to estimate the 90th percentile of this distribution with reasonable confidence. Using the six blanks available for this study, only the lower 60th to 70th percentile of this distribution can be estimated; therefore, this approach could underestimate the extent of contamination in environmental samples. In the present study, the most conservative approach was used, which assumes that contamination identified in any field or trip blank could occur in all environmental samples collected during the same sampling period. Although this approach can overestimate the extent of incidental contamination, no other procedure would ensure that this extent would not be underestimated. Therefore, detection of an analyte in any field or trip blank resulted in the censoring of concentrations of that analyte in all environmental samples collected during the same sampling period.

The results for 19 constituents in water were affected by censoring on the basis of contamination in laboratory, field, and trip blanks, as shown in table 15. Nine organic compounds and two trace elements were left with no detections in either sampling period after blank-censoring. Four additional organic compounds were left with no detections in the pre-landfall period; benzene and ammonia plus organic nitrogen were left with no detections in the post-landfall period. Four other constituents were censored to some extent, although some results still were quantified; two of these constituents were left with only one quantified value during the post-landfall period. Overall, 236 results out of a total of 1,189 results for the 19 constituents in table 15 were censored because of contamination in laboratory blanks (49 results) or field and trip blanks (187 results); however, 80 percent of these censored results were for only 5 constituents: toluene, ammonia plus organic nitrogen, mercury, organic carbon, and phosphorus.

Determination of Common Censoring Thresholds

Although the PPW test can be used with data censored at multiple reporting levels, it requires that the different reporting levels be randomly distributed between the two sample groups being compared. In this study, however, there were systematic differences in reporting levels between pre-landfall and post-landfall samples, especially for analytes that were determined by using different methods, by different laboratories, or both, for the two sampling periods (appendixes 1, 2). Therefore, all data for a given contaminant were censored to an "optimal" censoring threshold prior to statistical analysis, which is described in the next paragraph. For example, acenaphthene in sediment has an optimal censoring threshold of 0.36 micrograms per kilogram (μg/kg). Reported concentrations of 0.4, 0.2, and less than 1 μg/kg would be equivalent after censoring, respectively, to 0.4 μg/kg, less than 0.36 μg/kg, and indeterminate, which is defined in the next paragraph. Two-sided PPW tests were performed, and the sign of the test statistic indicated whether pre-landfall concentrations were higher than post-landfall concentrations or vice versa.

An optimal censoring threshold was computed for each analyte for which data were censored for one or more of the 96 samples in the paired-sample dataset, which consists of primary samples for sites sampled during both the pre-landfall and post-landfall periods. Many analytes had a wide range of reporting levels—one to three orders of magnitude. Selection of an optimal censoring threshold balanced two competing objectives: to include as many quantified detections as possible, but also to minimize the number of "indeterminate" samples. An indeterminate sample is defined as a sample with censored data—that is, reported less than a specified reporting level—for which the specified reporting level is higher than the applied censoring threshold,

so the sample cannot be classified as either a detection or nondetection at that threshold. As an example, censored data for acenaphthene in sediment ranged from less than 0.2 to less than 19 μg/kg, and quantified detections ranged from 0.34 to 2.1 μg/kg. If acenaphthene data are censored at the lowest possible censoring threshold of 0.2, then any censored value with a higher reporting threshold (for example, from less than 0.22 to less than 19 μg/kg) must be considered as indeterminate because we do not know whether the acenaphthene concentration is less than 0.2 or greater than or equal to 0.2 μg/kg. On the other hand, if we censor at the highest threshold of 19 μg/kg, then all samples with a detected concentration less than 19 μg/kg—in this case, all of the reported detections—become censored, reported as less than 19 μg/kg. The optimal censoring threshold was operationally defined as the lowest censoring level that converted no more than 5 percent of results from censored to indeterminate values, maximized the number of quantifiable detections, and if possible also minimized the number of indeterminate values. Because the optimal censoring threshold was designed for comparison of pre-landfall to post-landfall samples, it was determined by using the paired-sample dataset. For practical reasons, the maximum limit allowed for indeterminate values was raised slightly for some analytes that were determined in substantially fewer than the 96 samples typical of the paired-sample dataset, because it was difficult to meet the 5 percent maximum indeterminate value requirement and still preserve detections. Therefore, up to 7 percent indeterminate values were allowed for trace and major elements in the less than 63-μm sediment fraction, for which there were only about 70 samples, and up to 8 percent for selected analytes measured only during one sampling period, for which there were up to 48 samples.

The procedure for calculating the optimal censoring threshold for comparison of pre-landfall to post-landfall samples for a given analyte is illustrated for acenaphthene in sediment in figure 3. The x-axis shows possible censoring threshold concentrations for acenaphthene, which consist of all the reporting levels for censored samples. For acenaphthene, there are 94 samples, of which nine are quantified values. All of the observed reporting levels, from 0.2 to 19 μg/kg, were considered as possible censoring thresholds for this analyte, and each is represented in figure 3 with a gray bar showing the percentage of quantified values that would be "detections" if data were censored at that censoring threshold, except for 19 μg/kg, which is off the x-axis scale. The blue bars represent the percentage of samples that would be indeterminate at that threshold concentration because their reporting levels exceed the censoring threshold. The highest censoring threshold at which all 9 quantified values would still be "detections" after censoring would be 0.34.

Table 15. Constituents analyzed in water that were affected by censoring because of contamination detected in laboratory, field, and trip blanks from the Deepwater Horizon oil spill, Gulf of Mexico, 2010.

[**Abbreviations**: BTEX, benzene, toluene, ethylbenzene, xylene and related compounds; CARB, carbon, including petroleum hydrocarbons and hexane-extractable oil and grease; NUTR, nutrient; PAH, polycyclic aromatic hydrocarbon; post, post-landfall; pre, pre-landfall; TME, trace and major elements; VOC, volatile organic compound; –, not fully censored]

Analyte	Chemical class	Sampling period	Samples	Results after laboratory-blank censoring		Results after field and trip-blank censoring			
				Quantified results	Detection frequency (percent)	Samples censored due to blanks	Quantified results	Detection frequency (percent)	100 percent censored[1]
Organic contaminants									
1,2,3,5-Tetramethyl-benzene	VOC	Pre	60	1	2	1	0	0	Yes[2]
1,2,3-Trimethyl-benzene	VOC	Pre	60	2	3	2	0	0	Yes[2]
1,2,4-Trimethyl-benzene	VOC	Pre	60	3	5	3	0	0	Yes[2]
Acetone	VOC	Pre	62	5	8	5	0	0	Yes[3]
Benzene	VOC/BTEX	Post	48	3	6	3	0	0	Post only
Dichloromethane	VOC	Pre	62	3	5	3	0	0	Yes
Dichloromethane	VOC	Post	48	4	8	4	0	0	Yes
Dissolved organic carbon	CARB	Pre	62	62	100	41	21	34	–
Diesel range organics	CARB	Post	48	6	13	5	1	2	–
Ethylbenzene	VOC/BTEX	Pre	63	3	5	3	0	0	Yes[3]
Naphthalene	PAH	Post	48	1	2	1	0	0	Yes[3]
Toluene	VOC/BTEX	Pre	63	15	24	15	0	0	Pre only
Trichloromethane	VOC	Pre	62	3	5	3	0	0	Pre only
Xylene, *meta* plus *para*	VOC/BTEX	Pre	60	4	7	4	0	0	Yes[2]
Xylene, *ortho*	VOC/BTEX	Pre	60	3	5	3	0	0	Yes[2]
Trace and major elements, and nutrients in water									
Ammonia plus organic nitrogen as N	NUTR	Post	48	[4]40	83	40	0	0	Post only
Copper	TME	Post	48	22	46	3	19	40	–
Mercury	TME	Post	48	23	48	23	0	0	Yes[2]
Phosphorus as P	NUTR	Post	48	[4]22	46	21	1	2	–
Phosphorus as P	NUTR	Pre	68	[4]53	78	0	53	78	–
Silver	TME	Pre	63	4	6	4	0	0	Yes[3]

[1] Lists analytes that are 100 percent censored (no detections remaining) after blank-censoring. Post only, detected in post-landfall samples; pre only, detected in pre-landfall samples.

[2] Data available only for one sampling period.

[3] Not detected (without blank-censoring) in the other sampling period.

[4] Because of laboratory-blank contamination, results were censored for ammonia plus organic nitrogen in 8 of 48 post-landfall samples, phosphorus in 26 of 48 post-landfall samples, and phosphorus in 15 of 68 pre-landfall samples.

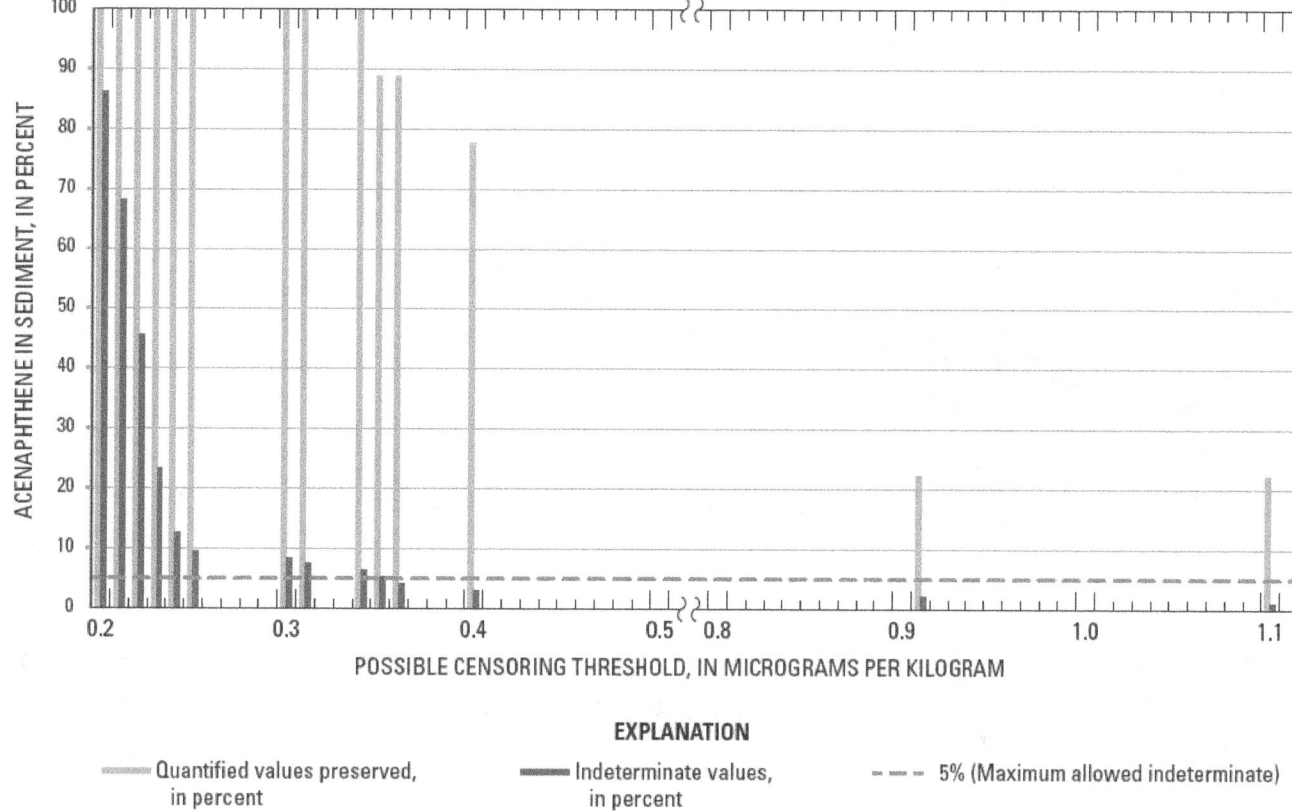

Figure 3. The effect of censoring threshold on the percentages of quantified values that are retained (gray bars) and indeterminate values (blue bars) for an example contaminant from the Deepwater Horizon oil spill, Gulf of Mexico, 2010: acenaphthene in sediment.

At a censoring threshold of 0.34, however, 6 percent of samples would be considered indeterminate because their reporting levels exceed 0.34. The maximum allowable limit for indeterminate samples is 5 percent, which is shown as the red line in figure 3, so a censoring threshold of 0.34 would not be acceptable. The lowest censoring threshold that meets the maximum indeterminate sample requirement is 0.35, which corresponds to 5 percent indeterminate samples. Raising the censoring threshold slightly to 0.36, however, would decrease the percentage of indeterminate samples slightly, to 4 percent, without censoring any quantified values. Increasing the censoring threshold again, such as to 0.40, would further reduce the indeterminate samples to 3 percent, but it also would result in loss of one more detection. The optimal censoring threshold selected was 0.36, which minimized the indeterminate samples and maximized quantifiable detections, while meeting the less than or equal to 5 percent criterion for maximum indeterminate samples.

Optimal censoring thresholds are shown in table 16 for individual analytes with at least 38 samples. Detection frequencies were calculated for each analyte at its optimal censoring threshold so that pre-landfall and post-landfall samples can be compared at a common detection threshold. Also, contaminant concentrations were censored at the optimal censoring threshold prior to statistical comparisons between sampling periods.

For all analytes of the same contaminant class and sampling medium, detection frequencies also were computed by using four common detection thresholds that allowed comparison among analytes with different MDLs. A range of common thresholds was used because the lower thresholds preserve more of the low-level quantified values, whereas higher thresholds allowed comparisons among a greater number of analytes. The four detection thresholds for a given contaminant class and sampling medium correspond to the 10th, 25th, 50th, and 70th percentiles in the distribution of optimal censoring thresholds for that contaminant type and sampling medium.

Table 16A. Optimal censoring thresholds and sample counts from the Deepwater Horizon oil spill, Gulf of Mexico, 2010: organic contaminants in water.

[**Abbreviations:** A, constituent was analyzed in samples from only one sampling period; B, constituent was not detected in any samples after blank censoring; BTEX, benzene, toluene, ethylbenzene, xylene and related compounds; C, constituent was not detected in paired-sample dataset; CARB, carbon, including petroleum hydrocarbons and hexane-extractable oil and grease; D, constituent was not detected in any samples after optimal censoring threshold was applied; F(x), field blank affected all samples collected during sampling period x; Lab, laboratory; na, not applicable; nc, not censored at optimal censoring threshold because no quantified detections remained after blank censoring; PAH, polycyclic aromatic hydrocarbon; post, post-landfall; PPW, paired Prentice-Wilcoxon; pre, pre-landfall; SVOC, semi-volatile organic compound (excluding PAHs); T(x), trip blank affected all samples collected during sampling period x; VOC, volatile organic compound; µg/L, microgram per liter; mg/L, milligram per liter; –, not detected in any blanks; <, less than]

Analyte	Chemical class	Type of blank with contamination (sampling period)[1]	Optimal censoring threshold[2]	Units	Before censoring — Number of samples with data	After censoring at optimal threshold — Number of quantified values that were censored	After censoring at optimal threshold — Number of censored values that are indeterminate[3]	PPW test performed	Reason no PPW test was run
1,2,3,5-Tetramethyl-benzene	VOC	F(pre)	nc	µg/L	38	na	na	No	A,B
1,2,3-Trimethylbenzene	VOC	F(pre)	nc	µg/L	38	na	na	No	A,B
1,2,4-Trimethylbenzene	VOC	F(pre)	nc	µg/L	38	na	na	No	A,B
1,3,5-Trimethylbenzene	VOC	–	0.064	µg/L	38	1	0	No	A
2,6-Dinitrotoluene	VOC	–	2.3	µg/L	92	0	4	No	C
2-Ethyltoluene	VOC	–	0.064	µg/L	38	1	0	No	A
Acenaphthene	PAH	–	0.28	µg/L	93	1	0	No	D
Acenaphthylene	PAH	–	0.3	µg/L	93	1	0	No	D
Acetone	VOC	F(pre)	nc	µg/L	87	na	na	No	B
Anthracene	PAH	–	0.39	µg/L	93	1	0	No	D
Benzene	VOC/BTEX	T(post)	0.34	µg/L	88	3	3	No	D
Benzo[a]anthracene	PAH	–	0.26	µg/L	93	1	0	Yes	na
Benzo[a]pyrene	PAH	–	0.33	µg/L	93	1	0	Yes	na
Benzo[b]fluoranthene	PAH	–	0.3	µg/L	93	2	0	Yes	na
Benzo[ghi]perylene	PAH	–	0.38	µg/L	43	0	0	No	A
Benzo[k]fluoranthene	PAH	–	0.4	µg/L	92	0	4	No	C
Bis(2-chloroethyl)ether	SVOC	–	2.1	µg/L	92	0	4	No	C
Bis(2-ethylhexyl) phthalate	SVOC	–	7.4	µg/L	93	1	4	Yes	na
Carbon disulfide	VOC	–	0.5	µg/L	84	1	0	Yes	na
Chloromethane	VOC	–	0.53	µg/L	85	1	0	No	D
Chrysene	PAH	–	0.33	µg/L	93	2	1	Yes	na
Dibenzo[a,h]anthracene	PAH	–	0.42	µg/L	92	0	0	No	C
Dibromomethane	VOC	–	0.1	µg/L	38	0	0	No	A,C
Dichloromethane	VOC	F(pre), T(post)	nc	µg/L	88	na	na	No	B
Diesel range organics	CARB	F(post)	73	µg/L	47	0	2	No	A
Diethyl phthalate	SVOC	F(post)	0.61	µg/L	93	1	0	No	D
Di-n-butyl phthalate	SVOC	–	2	µg/L	93	3	0	No	D
Di-n-octyl phthalate	SVOC	–	0.6	µg/L	92	0	0	No	C
Dissolved organic carbon	CARB	Lab	3	mg/L	87	25	4	Yes	na
Ethylbenzene	VOC/BTEX	F(pre)	nc	µg/L	88	na	na	No	B
Fluoranthene	PAH	–	0.3	µg/L	93	3	0	No	D
Fluorene	PAH	–	0.33	µg/L	93	1	0	No	D

Table 16A. Optimal censoring thresholds and sample counts from the Deepwater Horizon oil spill, Gulf of Mexico, 2010: organic contaminants in water.—Continued

[**Abbreviations**: A, constituent was analyzed in samples from only one sampling period; B, constituent was not detected in any samples after blank censoring; BTEX, benzene, toluene, ethylbenzene, xylene and related compounds; C, constituent was not detected in paired-sample dataset; CARB, carbon, including petroleum hydrocarbons and hexane-extractable oil and grease; D, constituent was not detected in any samples after optimal censoring threshold was applied; F(x), field blank affected all samples collected during sampling period x; Lab, laboratory; na, not applicable; nc, not censored at optimal censoring threshold because no quantified detections remained after blank censoring; PAH, polycyclic aromatic hydrocarbon; post, post-landfall; PPW, paired Prentice-Wilcoxon; pre, pre-landfall; SVOC, semi-volatile organic compound (excluding PAHs); T(x), trip blank affected all samples collected during sampling period x; VOC, volatile organic compound; µg/L, microgram per liter; mg/L, milligram per liter; –, not detected in any blanks; <, less than]

Analyte	Chemical class	Type of blank with contamination (sampling period)[1]	Optimal censoring threshold[2]	Units	Before censoring Number of samples with data	After censoring at optimal threshold Number of quantified values that were censored	Number of censored values that are indeterminate[3]	PPW test performed	Reason no PPW test was run
Gasoline range organics (C6-C10)	CARB	–	50	µg/L	49	1	0	No	A
Indeno(1,2,3-cd)pyrene	PAH	–	0.38	µg/L	92	0	0	No	C
Isophorone	SVOC	–	0.61	µg/L	93	10	5	No	D
Naphthalene	PAH	F(pre, post)	nc	µg/L	94	na	na	No	B
n-Propylbenzene	VOC	–	0.072	µg/L	38	0	0	No	A
Oil range organics (C28-C35)	CARB	–	46	µg/L	47	0	4	No	A
Organics (C8-C36)	CARB	–	47	µg/L	47	0	3	No	A
Pentachlorophenol	SVOC	–	3.1	µg/L	93	1	1	No	D
Phenanthrene	PAH	–	0.32	µg/L	93	1	0	No	D
Phenol	SVOC	–	1.5	µg/L	93	2	1	No	D
Pyrene	PAH	–	0.35	µg/L	93	3	0	No	D
Toluene	VOC/BTEX	F(pre)	0.7	µg/L	88	0	0	Yes	na
Tribromomethane	VOC	–	0.6	µg/L	88	1	0	No	D
Trichloromethane	VOC	F(pre)	0.6	µg/L	88	0	0	Yes	na
Xylene, *ortho*	VOC/BTEX	F(pre)	nc	µg/L	39	na	na	No	A,B
Xylenes, *meta* and *para*	VOC/BTEX	F(pre)	nc	µg/L	39	na	na	No	A,B
Xylenes, total	VOC/BTEX	–	1.6	µg/L	88	1	0	Yes	na

Table 16B. Optimal censoring thresholds and sample counts from the Deepwater Horizon oil spill, Gulf of Mexico, 2010: organic contaminants in whole sediment.

[Abbreviations: A, constituent was analyzed in samples from only one sampling period; B, constituent was not detected in any samples after optimal censoring threshold was applied; CARB, carbon, including petroleum hydrocarbons and hexane-extractable oil and grease; Lab, laboratory; na, not applicable; PAH, polycyclic aromatic hydrocarbon; PPW, paired Prentice-Wilcoxon; SVOC, semi-volatile organic compound (excluding PAHs); µg/kg, microgram per kilogram; mg/kg, milligram per kilogram; −, not detected in any blanks; <, less than]

Analyte	Chemical class	Type of blank with contamination (sampling period)[1]	Optimal censoring threshold[2]	Units	Before censoring — Number of samples with data	After censoring at optimal threshold — Number of quantified values that were censored	Number of censored values that are indeterminate[3]	PPW test performed	Reason no PPW test was run
1,2-Dimethyl-naphthalene	PAH	—	90	µg/kg	48	1	4	No	A
1,6-Dimethyl-naphthalene	SVOC	—	90	µg/kg	48	1	4	No	A
1-Methylnaphthalene	PAH	—	0.4	µg/kg	93	2	4	Yes	na
1-Methylphenanthrene	PAH	—	0.38	µg/kg	94	1	2	Yes	na
1-Methylpyrene	PAH	—	90	µg/kg	48	1	4	No	A
2,3,5-Trimethyl-naphthalene	PAH	—	0.27	µg/kg	47	0	3	No	A
2,3,6-Trimethyl-naphthalene	PAH	—	90	µg/kg	48	1	4	No	A
2,6-Dimethyl-naphthalene	PAH	—	0.47	µg/kg	94	1	2	Yes	na
2-Ethylnaphthalene	PAH	—	90	µg/kg	48	2	4	No	A
2-Methylnaphthalene	PAH	—	0.53	µg/kg	93	2	2	Yes	na
4H-Cyclopenta[d,e,f]-phenanthrene	PAH	—	90	µg/kg	48	2	4	No	A
9,10-Anthraquinone	SVOC	—	80	µg/kg	45	3	4	No	A
Acenaphthene	PAH	—	0.36	µg/kg	94	1	4	Yes	na
Acenaphthylene	PAH	—	0.38	µg/kg	94	2	2	Yes	na
Anthracene	PAH	—	0.38	µg/kg	94	0	1	Yes	na
Benzo[a]anthracene	PAH	—	0.25	µg/kg	94	0	3	Yes	na
Benzo[a]pyrene	PAH	—	0.25	µg/kg	94	2	3	Yes	na
Benzo[b]fluoranthene	PAH	—	0.23	µg/kg	94	1	0	Yes	na
Benzo[e]pyrene	PAH	—	0.38	µg/kg	94	0	2	Yes	na
Benzo[ghi]perylene	PAH	Lab	0.25	µg/kg	94	0	3	Yes	na
Benzo[k]fluoranthene	PAH	—	0.4	µg/kg	94	1	1	Yes	na
Biphenyl	SVOC	—	0.47	µg/kg	93	0	2	Yes	na
Bis(2-ethylhexyl) phthalate	SVOC	—	80	µg/kg	94	0	4	Yes	na
C1 Chrysene	PAH	—	1.5	µg/kg	93	0	1	Yes	na
C1 Dibenzothiophenes	PAH	—	1	µg/kg	93	0	4	Yes	na
C1 Fluoranthenes/pyrenes	PAH	—	0.91	µg/kg	93	0	4	Yes	na
C1 Fluorenes	PAH	—	1.5	µg/kg	93	0	2	Yes	na
C1 Naphthalenes	PAH	—	1.3	µg/kg	93	0	4	Yes	na
C1 Phenanthrenes/anthracenes	PAH	—	1.5	µg/kg	93	2	1	Yes	na
C2 Chrysenes	PAH	—	1	µg/kg	93	0	4	Yes	na
C2 Dibenzothiophenes	PAH	—	1	µg/kg	93	2	4	Yes	na
C2 Fluoranthenes/pyrenes	PAH	—	1.3	µg/kg	93	2	3	Yes	na
C2 Fluorenes	PAH	—	1.3	µg/kg	93	0	4	Yes	na
C2 Naphthalenes	PAH	—	0.91	µg/kg	93	0	2	Yes	na
C2 Phenanthrenes/anthracenes	PAH	—	1	µg/kg	93	0	4	Yes	na

Results and Discussion   45

Table 16B. Optimal censoring thresholds and sample counts from the Deepwater Horizon oil spill, Gulf of Mexico, 2010: organic contaminants in whole sediment.—Continued

[Abbreviations: A, constituent was analyzed in samples from only one sampling period; B, constituent was not detected in any samples after optimal censoring threshold was applied; CARB, carbon, including petroleum hydrocarbons and hexane-extractable oil and grease; Lab, laboratory; na, not applicable; PAH, polycyclic aromatic hydrocarbon; PPW, paired Prentice-Wilcoxon; SVOC, semi-volatile organic compound (excluding PAHs); µg/kg, microgram per kilogram; mg/kg, milligram per kilogram; <, less than]

Analyte	Chemical class	Type of blank with contamination (sampling period)[1]	Optimal censoring threshold[2]	Units	Before censoring Number of samples with data	After censoring at optimal threshold Number of quantified values that were censored	Number of censored values that are indeterminate[3]	PPW test performed	Reason no PPW test was run
C3 Chrysenes	PAH	–	1	µg/kg	93	0	4	Yes	na
C3 Dibenzothiophenes	PAH	–	1	µg/kg	93	1	4	Yes	na
C3 Fluoranthenes/pyrenes	PAH	–	1.5	µg/kg	93	0	1	Yes	na
C3 Fluorenes	PAH	–	1.8	µg/kg	93	0	1	Yes	na
C3 Naphthalenes	PAH	–	0.91	µg/kg	93	0	2	Yes	na
C3 Phenanthrenes/qnthracenes	PAH	–	1	µg/kg	93	1	4	Yes	na
C4 Chrysenes	PAH	–	1.3	µg/kg	93	2	4	Yes	na
C4 Dibenzothiophenes	PAH	–	1.8	µg/kg	93	3	1	Yes	na
C4 Naphthalenes	PAH	–	1.2	µg/kg	93	0	4	Yes	na
C4 Phenanthrenes/anthracenes	PAH	–	1.3	µg/kg	93	0	3	Yes	na
Carbazole	SVOC	–	80	µg/kg	91	4	5	No	B
Chrysene	PAH	–	0.23	µg/kg	94	1	0	Yes	na
Dibenzo[ah]anthracene	PAH	–	0.38	µg/kg	94	1	3	Yes	na
Dibenzothiophene	SVOC	–	0.38	µg/kg	94	1	2	Yes	na
Fluoranthene	PAH	–	0.26	µg/kg	94	0	0	Yes	na
Fluorene	PAH	–	0.44	µg/kg	94	1	3	Yes	na
Indeno[123cd]pyrene	PAH	Lab	0.23	µg/kg	94	0	1	Yes	na
Naphthalene	PAH	–	0.29	µg/kg	94	6	5	Yes	na
Oil and grease	CARB	Lab	110	mg/kg	94	16	2	Yes	na
Perylene	PAH	Lab	0.22	µg/kg	94	0	5	Yes	na
Petroleum hydrocarbons	CARB	–	340	mg/kg	46	1	4	No	na
Phenanthrene	PAH	–	0.23	µg/kg	94	2	2	Yes	na
Pyrene	PAH	–	0.24	µg/kg	94	0	1	Yes	na
Total organic carbon	CARB	–	0.1	percent	93	0	0	Yes	na
Total organic carbon	CARB	–	1100	mg/kg	46	0	4	No	A

[1]In affected samples, reported concentrations were censored at five times the maximum concentration detected in the corresponding laboratory blank.

[2]Lowest censoring level that maximizes the number of quantifiable detections, minimizes the number of indeterminate samples, and for which the percent indeterminate samples is no more than 5 to 8 percent, depending on sample size.

[3]Indeterminate samples are censored data for which the reporting level (for example, <1) is higher than the applied censoring threshold (for example, 0.2), so it cannot be classified as either a detection or nondetection at that threshold (for example, it is unknown whether the contaminant is present at levels above 0.2).

Table 16C.   Optimal censoring thresholds and sample counts from the Deepwater Horizon oil spill, Gulf of Mexico, 2010: trace and major elements and nutrients in water.

[**Abbreviations**: A, constituent was not detected in any samples after optimal censoring threshold was applied; B, constituent was analyzed in samples from only one sampling period; C, constituent was not detected in any samples after blank censoring; F(x), field blank affected all samples collected during sampling period x; Lab, laboratory; na, not applicable; nc, data not censored; NUTR, nutrient; post, post-landfall; PPW, paired Prentice-Wilcoxon; pre, pre-landfall; TME, trace and major element; µg/L, microgram per liter; mg/L, milligram per liter; na, not applicable; nc, not censored; –, not detected in any blanks; <, less than]

Analyte	Symbol or abbreviation	Chemical class	Type of blank with contamination (sampling period)[1]	Optimal censoring threshold[2]	Units	Before censoring — Number of samples with data	After censoring at optimal threshold — Number of quantified values that were censored	Number of censored values that are indeterminate[3]	PPW test performed	Reason no PPW test was run
Aluminum	Al	TME	–	400	µg/L	88	12	3	Yes	na
Ammonia as N	N (ammonia)	NUTR	–	0.04	mg/L	91	14	0	Yes	na
Ammonia as NH$_4$	N (ammonium)	NUTR	–	0.0515	mg/L	91	14	0	Yes	na
Ammonia plus organic nitrogen	N (Kjeldahl)	NUTR	Lab, F(post)	2.4	mg/L	93	46	5	No	A
Arsenic	As	TME	F(pre)	40	µg/L	80	39	0	No	A
Barium	Ba	TME	–	nc[4]	µg/L	80	na	na	Yes	na
Beryllium	Be	TME	–	10	µg/L	80	4	0	No	A
Boron	B	TME	Lab, F(post)	nc[4]	µg/L	42	na	na	No	B
Cadmium	Cd	TME	–	10	µg/L	80	3	0	No	A
Calcium	Ca	TME	Lab, F(pre, post)	nc[4]	mg/L	86	na	na	Yes	na
Chromium	Cr	TME	–	20	µg/L	80	20	0	Yes	na
Cobalt	Co	TME	–	30	µg/L	80	26	0	No	A
Copper	Cu	TME	F(post)	38	µg/L	80	17	0	No	A
Dissolved nitrogen	N (total)	NUTR	–	nc[4]	mg/L	86	na	na	Yes	na
Iron	Fe	TME	–	500	µg/L	80	17	0	Yes	na
Lead	Pb	TME	–	20	µg/L	80	20	0	Yes	na
Lithium	Li	TME	F(pre)	nc[4]	µg/L	39	na	na	No	B
Magnesium	Mg	TME	Lab, F(pre, post)	nc[4]	mg/L	86	na	na	Yes	na
Manganese	Mn	TME	–	10	µg/L	80	5	5	Yes	na
Mercury	Hg	TME	F(post)	nc[5]	µg/L	40	na	na	No	B,C
Molybdenum	Mo	TME	–	20	µg/L	80	52	0	Yes	na
Nickel	Ni	TME	–	75	µg/L	80	22	0	No	A
Organic nitrogen	N (organic)	NUTR	–	2.4	mg/L	91	33	4	Yes	na
Phosphorus as P	P	NUTR	Lab, F(post)	0.18	mg/L	94	29	4	Yes	na

Table 16C. Optimal censoring thresholds and sample counts from the Deepwater Horizon oil spill, Gulf of Mexico, 2010: trace and major elements and nutrients in water. —Continued

[**Abbreviations**: A, constituent was not detected in any samples after optimal censoring threshold was applied; B, constituent was analyzed in samples from only one sampling period; C, constituent was not detected in any samples after blank censoring; F(x), field blank affected all samples collected during sampling period x; Lab, laboratory; na, not applicable; nc, data not censored; NUTR, nutrient; post, post-landfall; PPW, paired Prentice-Wilcoxon; pre, pre-landfall; TME, trace and major element; μg/L, microgram per liter; mg/L, milligram per liter; na, not applicable; nc, not censored; –, not detected in any blanks; <, less than]

Analyte	Symbol or abbreviation	Chemical class	Type of blank with contamination (sampling period)[1]	Optimal censoring threshold[2]	Units	Before censoring — Number of samples with data	After censoring at optimal threshold — Number of quantified values that were censored	Number of censored values that are indeterminate[3]	PPW test performed	Reason no PPW test was run
Potassium	K	TME	Lab, F(post)	nc[4]	mg/L	86	na	na	Yes	na
Selenium	Se	TME	–	40	μg/L	80	15	0	No	A
Silver	Ag	TME	F(pre)	nc[5]	μg/L	80	na	na	No	C
Sodium	Na	TME	Lab, F(pre, post)	nc[4]	mg/L	86	na	na	Yes	na
Strontium	Sr	TME	–	nc[4]	μg/L	39	na	na	No	B
Vanadium	V	TME	–	20	μg/L	42	na	na	No	B
Zinc	Zn	TME	–	80	μg/L	80	12	0	Yes	na

[1]For constituents detected in one or more laboratory-, field- or trip-blank samples, reported concentrations were censored at five times the concentration in the corresponding laboratory blank, or at five times the maximum concentration detected in field or trip blanks for that sampling period.

[2]Lowest detection threshold that maximizes the number of quantifiable detections, minimizes the number of indeterminate samples, and for which the percent indeterminate samples is no more than 5 to 8 percent, depending on sample size.

[3]Indeterminate samples are censored data for which the reporting level (for example, <1) is higher than the applied censoring threshold (for example, 0.2), so it cannot be classified as either a detection or nondetection at that threshold (for example, it is unknown whether the contaminant is present at levels above 0.2).

[4]Data not censored because constituent was detected in all samples.

[5]Optimal censoring threshold was not applied because no quantified detections remained after blank censoring.

Table 16D. Optimal censoring thresholds and sample counts from the Deepwater Horizon oil spill, Gulf of Mexico, 2010: trace and major elements and nutrients in whole sediment.

[**Abbreviations**: CARB, carbon; mg/kg, milligram per kilogram; na, not applicable; nc, not censored because constituent was detected in all samples; NUTR, nutrient; PHYS, physical property; PLT63, percent of sediment sample that passes through a 63-μm sieve; PPW, paired Prentice-Wilcoxon; TME, trace and major element; <, less than; μm, micrometer]

Analyte	Symbol or abbreviation	Chemical class	Optimal censoring threshold PPW test[1]	Units	Before censoring Number of samples with data	After censoring at optimal threshold Number of quantified values that were censored	Number of censored values that are indeterminate[2]	PPW test performed
Aluminum	Al	TME	0.1	percent	96	0	2	Yes
Antimony	Sb	TME	0.1	mg/kg	96	0	2	Yes
Arsenic	As	TME	0.1	mg/kg	96	0	0	Yes
Barium	Ba	TME	1	mg/kg	96	0	2	Yes
Beryllium	Be	TME	0.1	mg/kg	96	0	2	Yes
Cadmium	Cd	TME	0.1	mg/kg	96	0	3	Yes
Calcium	Ca	TME	0.1	mg/kg	96	0	2	Yes
Carbon, total	TC	CARB	0.1	percent	96	0	0	Yes
Chromium	Cr	TME	2	mg/kg	96	0	0	Yes
Cobalt	Co	TME	1	mg/kg	96	0	2	Yes
Copper	Cu	TME	1	mg/kg	96	0	2	Yes
Iron	Fe	TME	0.1	percent	96	0	2	Yes
Lead	Pb	TME	1	mg/kg	96	0	0	Yes
Lithium	Li	TME	1	mg/kg	96	0	2	Yes
Magnesium	Mg	TME	0.1	percent	96	0	2	Yes
Manganese	Mn	TME	1	mg/kg	96	0	0	Yes
Mercury	Hg	TME	0.01	mg/kg	96	0	0	Yes
Molybdenum	Mo	TME	1	mg/kg	96	0	3	Yes
Nickel	Ni	TME	1	mg/kg	96	0	2	Yes
Nitrogen	N	NUTR	0.1	percent	95	20	0	Yes
Phosphorus	P	NUTR	nc	mg/kg	96	na	na	Yes
Potassium	K	TME	0.1	percent	96	0	2	Yes
Selenium	Se	TME	0.1	mg/kg	96	0	3	Yes
Silver	Ag	TME	0.5	mg/kg	96	0	3	Yes
Sodium	Na	TME	0.1	percent	96	0	0	Yes
Strontium	Sr	TME	1	mg/kg	96	0	0	Yes
Sulfur	S	TME	0.01	percent	96	0	0	Yes
Tin	Sn	TME	1	mg/kg	96	0	2	Yes
Titanium	Ti	TME	0.01	percent	96	0	1	Yes
Vanadium	V	TME	1	mg/kg	96	0	2	Yes
Zinc	Zn	TME	1	mg/kg	96	0	2	Yes
Percent of sediment <63-μm	PLT63	PHYS	1	percent	95	0	0	Yes

[1]Lowest detection threshold that maximizes the number of quantifiable detections, minimizes the number of indeterminate samples, and for which the percent indeterminate samples is no more than 5 to 8 percent, depending on sample size.

[2]Indeterminate samples are censored data for which the reporting level (for example, <1) is higher than the applied censoring threshold (for example, 0.2), so it cannot be classified as either a detection or nondetection at that threshold (it is unknown whether the contaminant is present at levels above 0.2).

Table 16E.   Optimal censoring thresholds and sample counts from the Deepwater Horizon oil spill, Gulf of Mexico, 2010: trace and major elements and nutrients in the less than 63-micrometer (μm) sediment fraction

[**Abbreviations**: A, constituent was not detected in any samples after optimal censoring threshold was applied; CARB, carbon; mg/kg, milligram per kilogram; na, not applicable; nc, not censored because constituent was detected in all samples; NUTR, nutrient; PHYS, physical property; PPW, paired Prentice-Wilcoxon; TME, trace and major element; <, less than]

Analyte or parameter	Symbol	Chemical class	Optimal censoring threshold[1]	Units	Before censoring	After censoring at optimal threshold		PPW test perfromed	Reason no PPW test was run
					Number of samples with data	Number of quantified values that were censored	Number of censored values that are indeterminate[2]		
Aluminum	Al	TME	0.3	percent	70	0	1	Yes	na
Antimony	Sb	TME	0.4	mg/kg	70	3	5	Yes	na
Arsenic	As	TME	nc	mg/kg	70	na	na	Yes	na
Barium	Ba	TME	nc	mg/kg	70	na	na	Yes	na
Beryllium	Be	TME	0.9	mg/kg	70	7	5	Yes	na
Cadmium	Cd	TME	1.3	mg/kg	70	42	3	No	A
Calcium	Ca	TME	0.2	percent	70	0	1	Yes	na
Carbon, total	TC	CARB	nc	percent	39	na	na	Yes	na
Chromium	Cr	TME	9	mg/kg	79	3	3	Yes	na
Cobalt	Co	TME	10	mg/kg	70	24	5	Yes	na
Copper	Cu	TME	5	mg/kg	79	0	1	Yes	na
Iron	Fe	TME	0.2	percent	70	0	1	Yes	na
Lead	Pb	TME	3	mg/kg	70	0	5	Yes	na
Lithium	Li	TME	7	mg/kg	70	0	1	Yes	na
Magnesium	Mg	TME	nc	percent	79	0	1	Yes	na
Manganese	Mn	TME	nc	mg/kg	70	na	na	Yes	na
Mercury	Hg	TME	0.01	mg/kg	47	0	0	Yes	na
Molybdenum	Mo	TME	13	mg/kg	70	20	5	Yes	na
Nickel	Ni	TME	2	mg/kg	70	0	1	Yes	na
Nitrogen	N	NUTR	nc	percent	39	na	na	Yes	na
Phosphorus	P	NUTR	1	mg/kg	79	0	0	Yes	na
Potassium	K	TME	0.6	percent	70	0	4	Yes	na
Selenium	Se	TME	1.2	mg/kg	70	45	5	Yes	na
Sodium	Na	TME	0.5	percent	70	0	2	Yes	na
Strontium	Sr	TME	nc	mg/kg	70	na	na	Yes	na
Sulfur	S	TME	nc	percent	70	na	na	Yes	na
Tin	Sn	TME	13	mg/kg	70	29	5	No	A
Titanium	Ti	TME	0.03	percent	70	0	4	Yes	na
Uranium	U	TME	600	mg/kg	70	1	5	No	A
Vanadium	V	TME	6	mg/kg	79	0	1	Yes	na
Zinc	Zn	TME	20	mg/kg	79	0	0	Yes	na

[1]Lowest detection threshold that maximizes the number of quantifiable detections, minimizes the number of indeterminate samples, and for which the percent indeterminate samples is no more than 7 percent.

[2]Indeterminate samples are censored data for which the reporting level (for example, <1) is higher than the applied censoring threshold (for example, 0.2), so it cannot be classified as either a detection or nondetection at that threshold (for example, it is unknown whether the contaminant is present at levels above 0.2).

Organic Contaminants in Water

For organic contaminants in water, samples were analyzed by different laboratories; pre-landfall samples were analyzed by the USGS NWQL, and post-landfall samples by the TestAmerica Laboratories in either Colorado or Florida. This complicates the comparison of contaminant occurrence between sampling periods, as described in the following section.

Contaminant Occurrence

Few organic contaminants were detected in water samples (table 17). For each contaminant, table 17 provides an optimal censoring threshold, as described previously, to use in comparing detection frequencies between pre-landfall and post-landfall samples, as well as a series of four common detection thresholds to use in comparing detection frequencies among analytes. A common detection threshold must be applied when comparing detection frequencies for analytes with different or variable reporting levels, as is discussed later in this report.

Of the 41 contaminants analyzed only in pre-landfall samples, where the number of samples (n) is 60 to 65 sites depending on the analyte, 5 contaminants were detected in one or more samples: 1,3,5-trimethylbenzene, 2-ethyltoluene, n-propylbenzene, dibromomethane, and benzo[g,h,i] perylene. Of the 24 contaminants analyzed only in post-landfall samples, where n is 48 sites, 5 contaminants were detected in one or more samples: a mixture of C8 to C36 organics, oil range organics (C28 to C35), gasoline-range organics (C6 to C10), diesel-range organics, and total xylene. Of 94 organic contaminants analyzed in both pre-landfall samples and post-landfall samples, one or more detections were observed for 28 analytes in pre-landfall samples and for 9 analytes in post-landfall samples, with 7 of these analytes, including dissolved organic carbon, detected in samples from both sampling periods. Two analytes—toluene and trichloromethane—were detected in one or more post-landfall samples but no pre-landfall samples. Although more analytes were detected in pre-landfall than post-landfall samples, two factors need to be considered: (1) more sites distributed over a wider geographic area were sampled during the pre-landfall period, typically 60 to 68, than during the post-landfall period, which typically had 47 to 48 sites; and (2) reporting levels were lower for many analytes in pre-landfall than in post-landfall samples, which were analyzed by different laboratories. Thus, the detection frequencies are not directly comparable without adjustment for these factors.

This is illustrated in figure 4, which shows the cumulative frequency distributions of concentrations determined for two example contaminants in water, isophorone and benzene. (Appendix 2 provides a complete set of cumulative frequency plots for all individual contaminants determined in water and sediment.) For isophorone in water (fig. 4A), the detections observed in many pre-landfall samples were well below the reporting level for isophorone in post-landfall samples. Although it is possible that isophorone was present in post-landfall samples at concentrations comparable to those in pre-landfall samples, the analytical method used for post-landfall samples was not sensitive enough to detect these values. Similar results were observed for several PAHs in water (appendix 2-1). The benzene example (fig. 4B) illustrates the effect of blank censoring. In this case, comparison of pre-landfall to post-landfall occurrence is limited because the censoring level for all post-landfall samples was raised to 2.1 µg/L as a result of benzene detection in a blank from the post-landfall period. Because the raw benzene concentrations detected in post-landfall samples were less than the censoring threshold, there is uncertainty as to whether these concentrations were the result of incidental contamination; therefore, all post-landfall samples were reported as less than 2.1 µg/L. Concentrations of 0.02 to 0.05 µg/L that were detected in pre-landfall samples were much lower than the less than 2.1 µg/L censored results for post-landfall samples, so pre-landfall and post-landfall sample concentrations cannot be compared quantitatively for this analyte.

When detection frequencies above the optimal censoring threshold, which varies by analyte, as shown in table 17, were computed for organic contaminants in water, dissolved organic carbon was detected in about 40 percent of samples from both pre-landfall and post-landfall sampling periods, and 14 additional analytes were detected in one or more samples. Of these 14 analytes, 12 were detected in only one sample each. The remaining two detected analytes were toluene and the mixture of C8 to C36 organics. Toluene was detected above an optimal censoring threshold of 0.7 µg/L in 13 percent of post-landfall samples and no pre-landfall samples; the C8 to C36 organics were detected above an optimal censoring threshold of 47 µg/L in 7 percent detection of post-landfall samples but were not analyzed in pre-landfall samples. Toluene is the only analyte of the 94 determined in water during both sampling periods to show much difference between the two sampling periods in detection frequencies above the optimal censoring threshold (table 17). A more rigorous, statistical comparison between contaminant concentrations in pre-landfall and post-landfall samples follows.

Table 17.   Summary statistics for organic contaminants in water from the Deepwater Horizon oil spill, Gulf of Mexico, 2010.

This table is available as a Microsoft© Excel spreadsheet. It can be accessed and downloaded at URL http://pubs.usgs.gov/ sir/2012/5228.

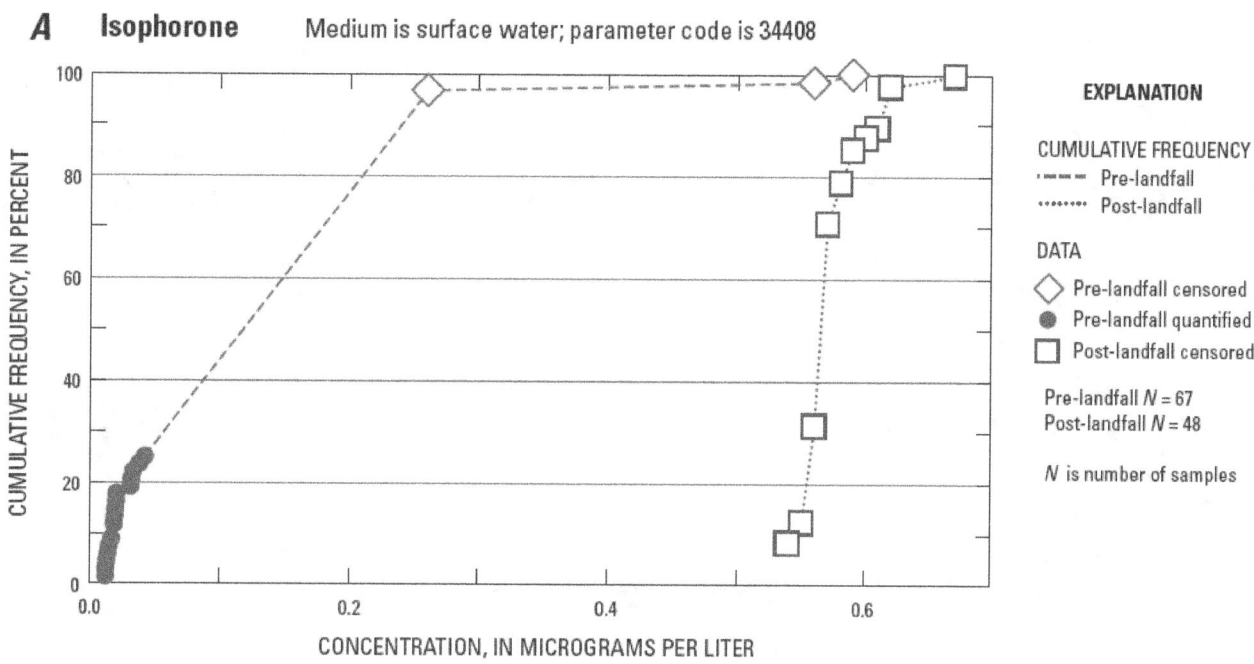

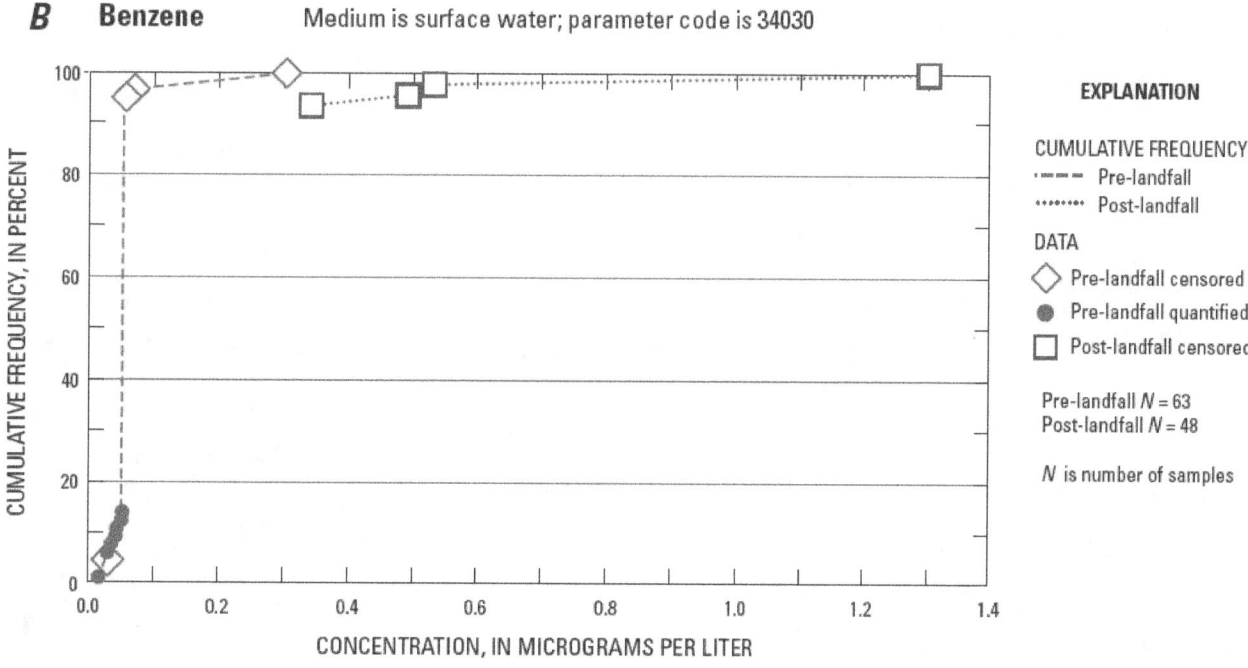

Figure 4. Examples of the data distribution of contaminant concentrations from the Deepwater Horizon oil spill, Gulf of Mexico, 2010: (*A*) isophorone in water, (*B*) benzene in water, (*C*) C3-alkylated fluorenes in sediment, (*D*) zinc in water, (*E*) molybdenum in water, (*F*) phosphorus in water, (*G*) ammonia plus organic nitrogen in water, (*H*) potassium in water, (*I*) calcium in whole sediment, (*J*) lead in whole sediment, and (*K*) phosphorus in whole sediment. N, number of samples.

C   **C3-alkylated fluorenes**   Medium is sediment; parameter code is 68091

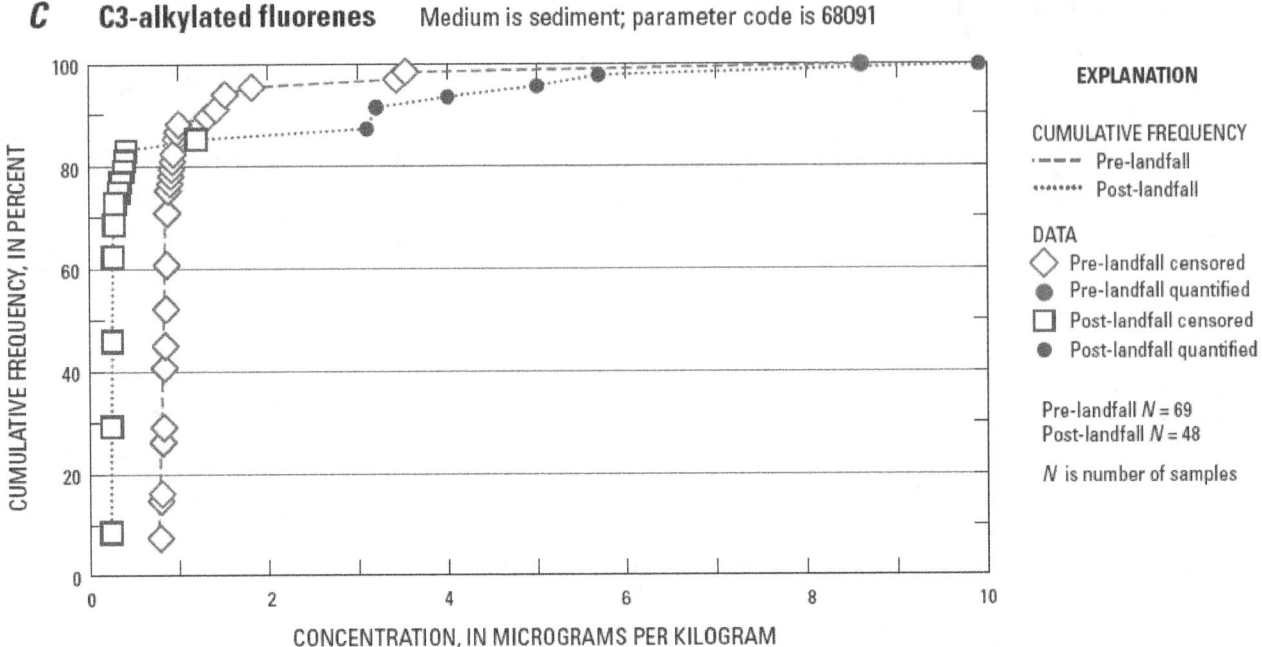

D   **Zinc**   Medium is surface water; parameter code is 01092

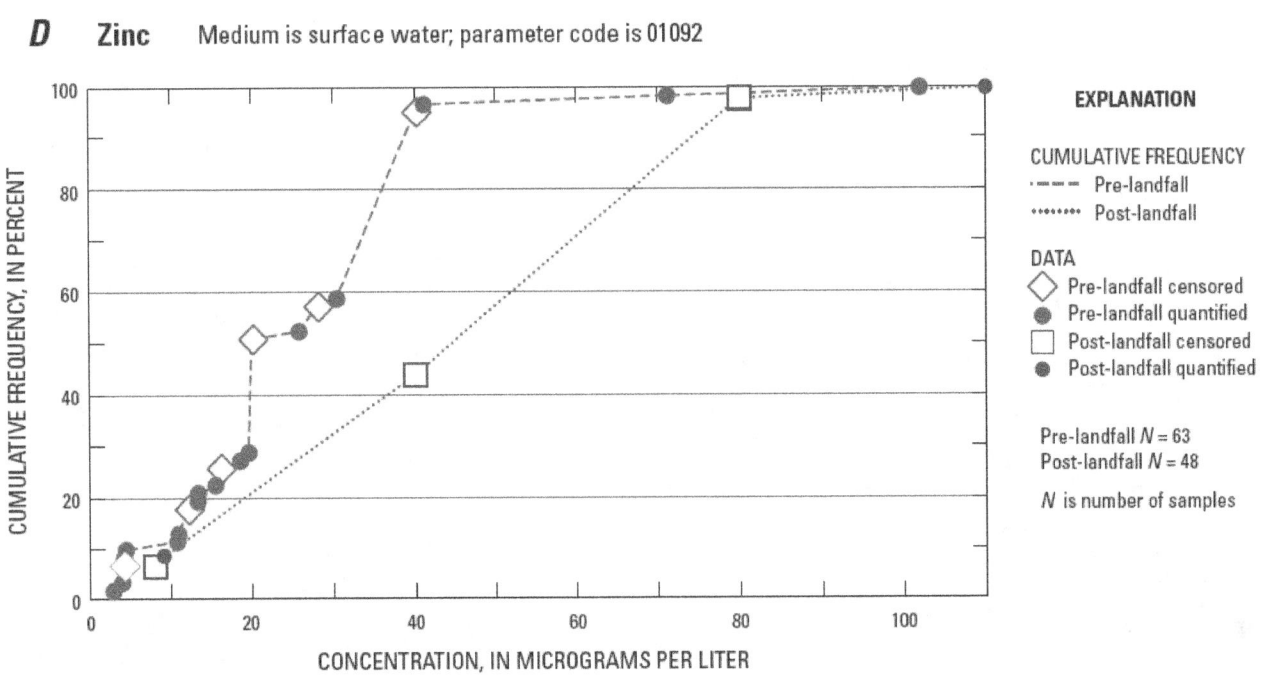

Figure 4.—Continued

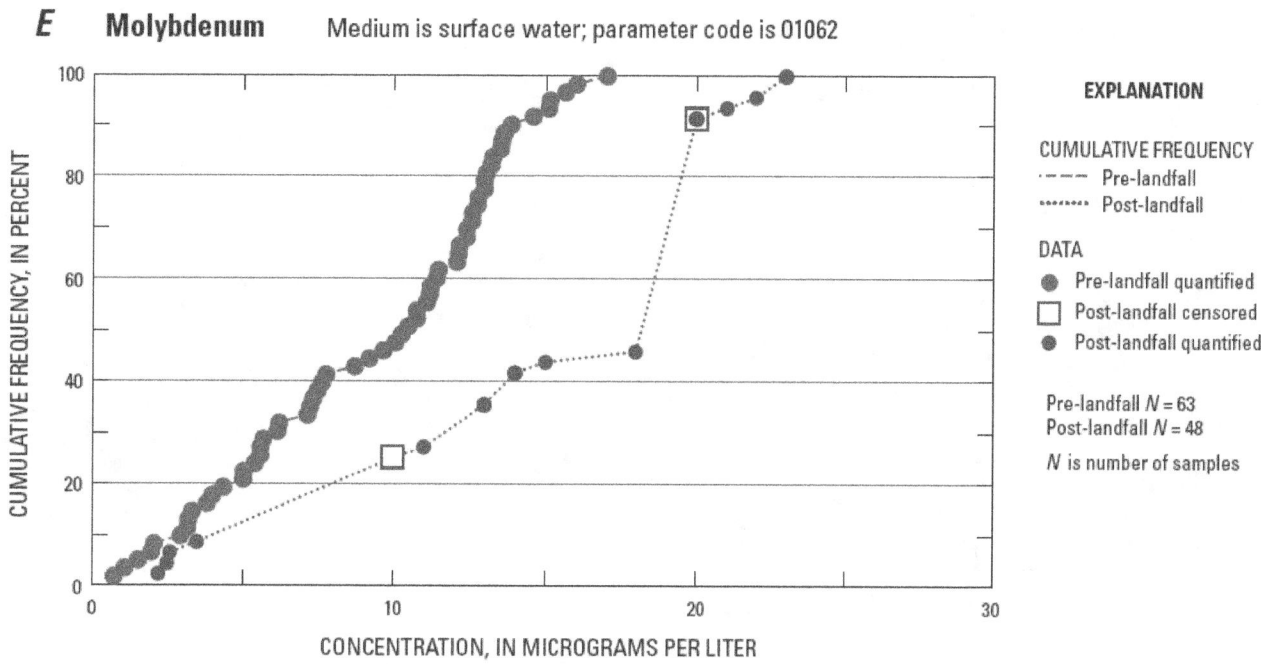

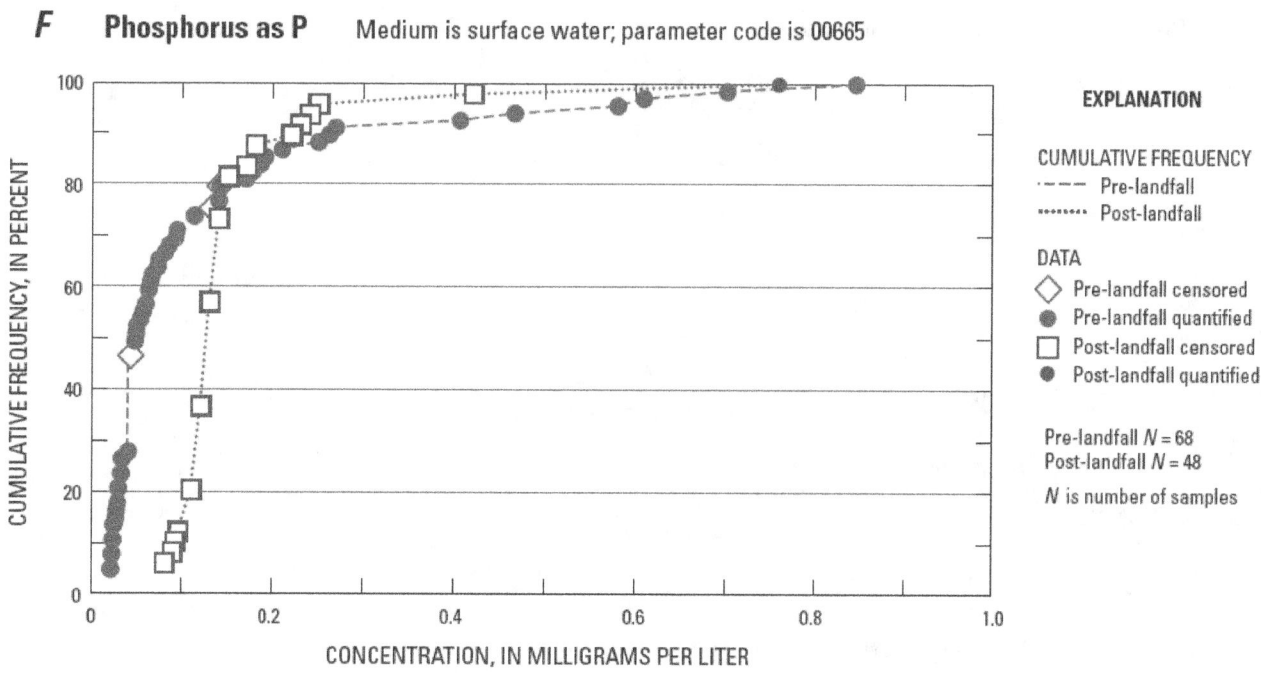

Figure 4.—Continued

G   **Ammonia plus organic nitrogen as N**   Medium is surface water; parameter code is 00625

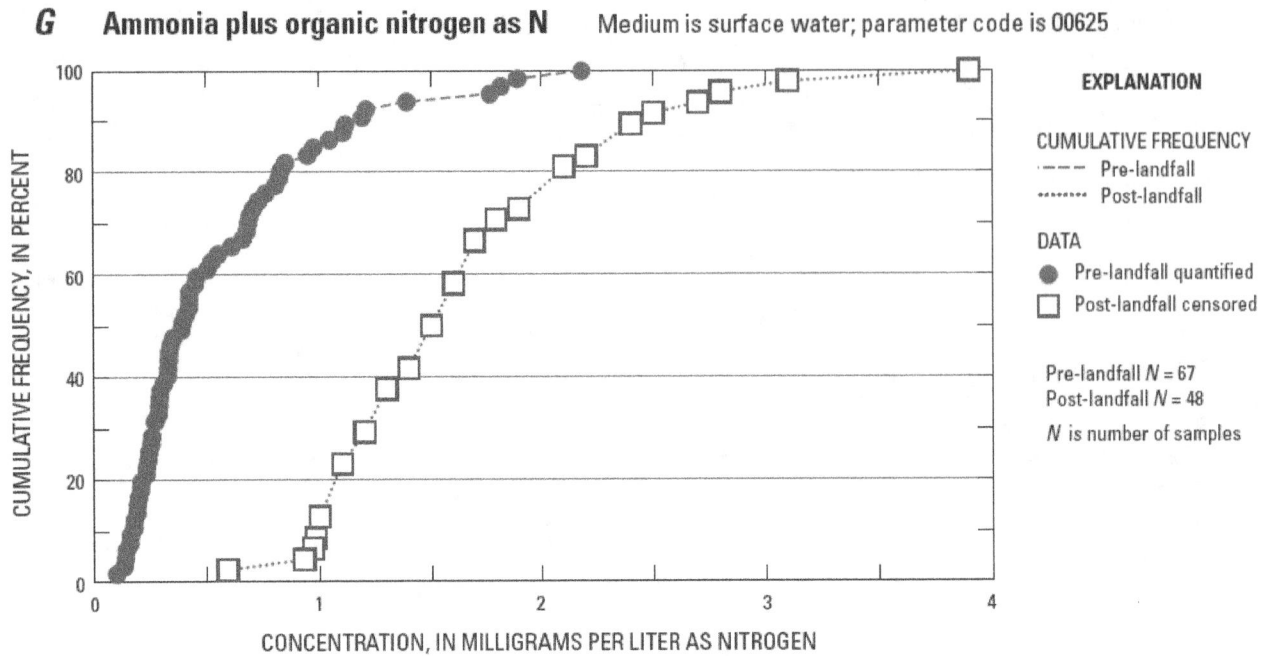

H   **Potassium**   Medium is surface water; parameter code is 00937

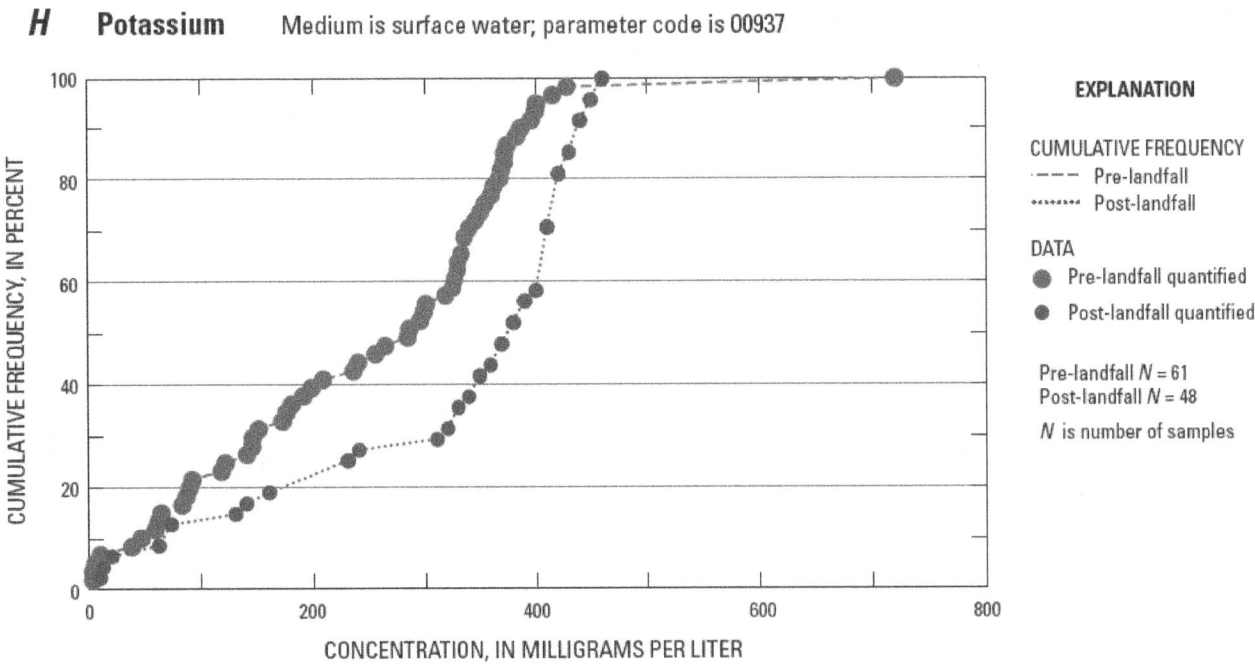

Figure 4.—Continued

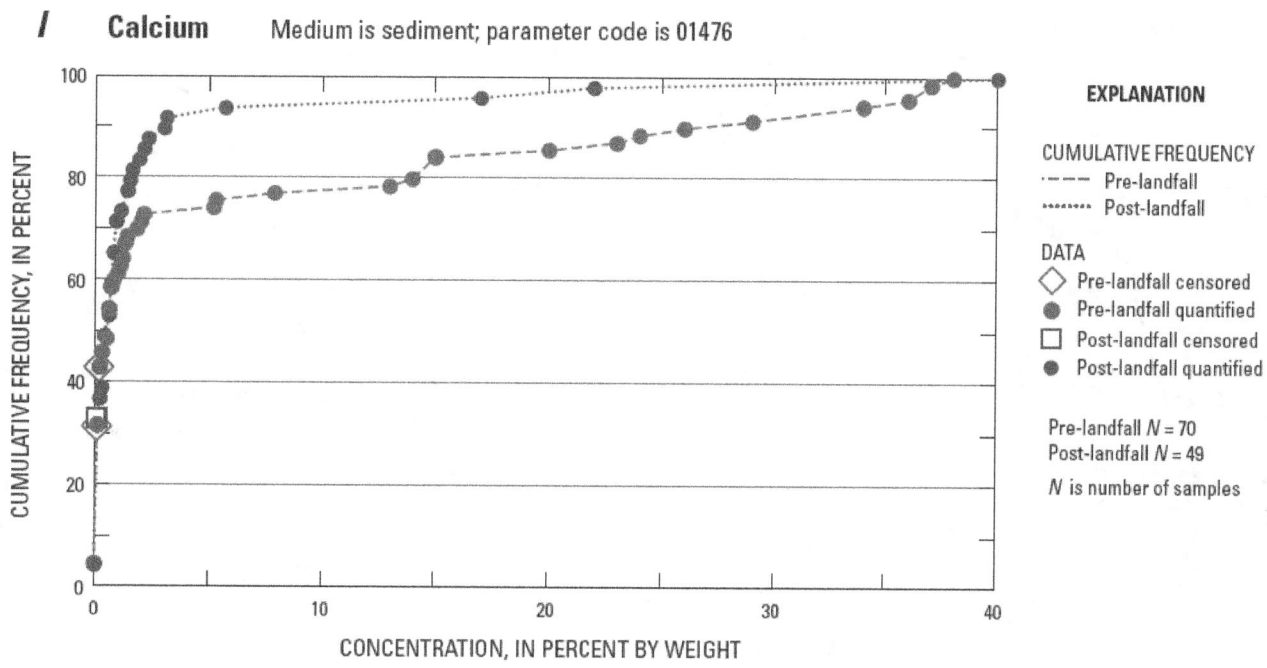

I   **Calcium**   Medium is sediment; parameter code is 01476

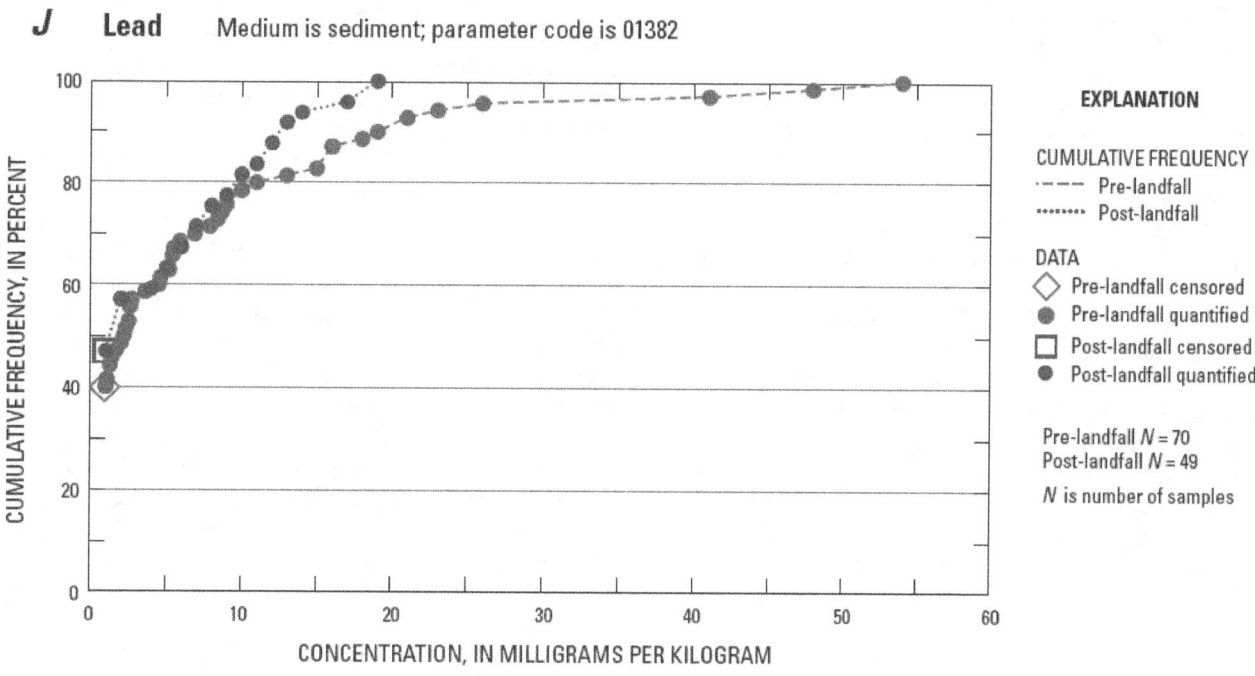

J   **Lead**   Medium is sediment; parameter code is 01382

Figure 4.—Continued

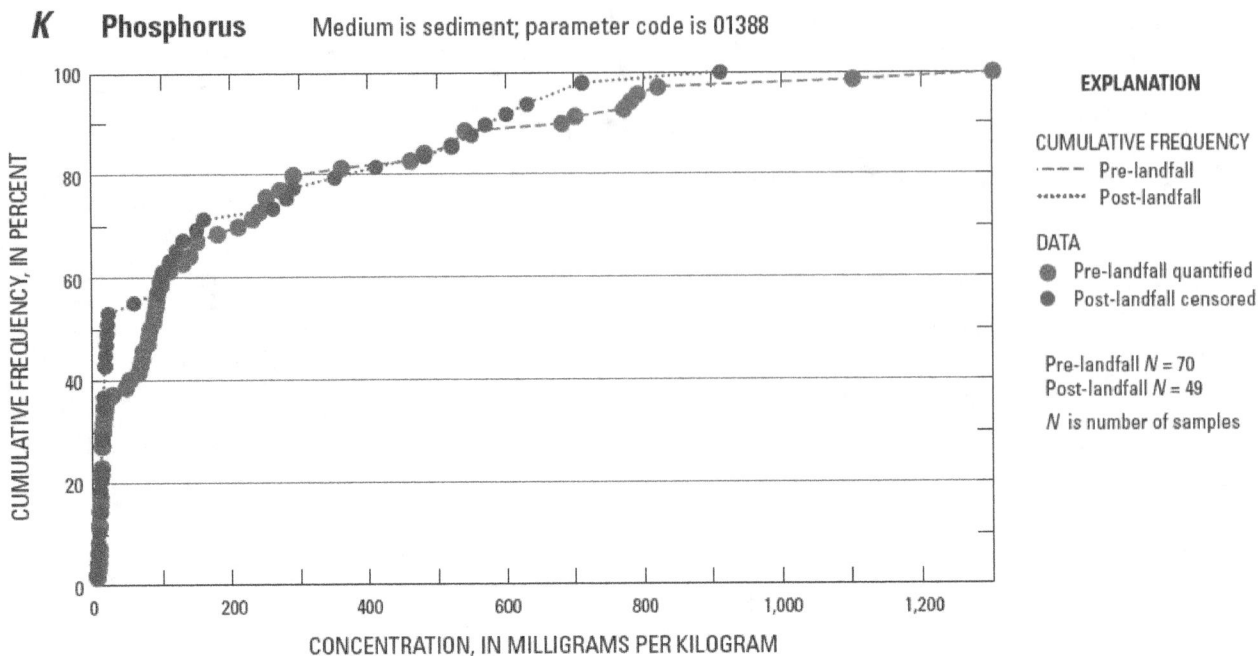

K   **Phosphorus**   Medium is sediment; parameter code is 01388

Figure 4.—Continued

Comparison of Pre-Landfall to Post-Landfall Samples

Of the approximately 100 organic compounds that were determined in at least 100 water samples, only 11 compounds had enough quantified detections above the optimal censoring threshold to make a statistical comparison of pre-landfall to post-landfall samples. Of these, only toluene and organic carbon showed a significant difference between pre-landfall and post-landfall samples in the PPW test. Toluene concentrations were significantly higher in post-landfall samples than in pre-landfall samples (p=0.0144; table 18). This statistical test result supports the previous observation that toluene in water had a higher detection frequency in post-landfall samples, at 13 percent, than in pre-landfall samples, where it was not detected, after data were censored to an optimal threshold of 0.7 µg/L (table 17). The difference in concentrations between post-landfall and pre-landfall samples ($C_{post} - C_{pre}$) for toluene at each site along the GOM coast, from west to east, is shown in figure 5A. For each site

in figure 5A, the difference in concentrations of toluene is a range, which indicates that one or both samples is censored, that is, a nondetection; this range is derived by using both zero and the reporting level as the censored value when calculating the difference. By using this method of calculation, all of the bars that are centered on zero are cases where both pre-landfall samples and post-landfall samples were censored (for example, most sites in fig. 5A); ranges that do not include zero are based on one censored value and one detection. For toluene (fig. 5A), the five bars with positive values indicate detections in post-landfall samples and censored data in the corresponding pre-landfall samples. If there is a single point instead of a range, then both samples were quantified detections. A single negative value results when the concentration is higher in the pre-landfall sample than in the post-landfall sample (as shown in fig. 5B for LA-22), and a single positive value indicates that the concentration is higher in the post-landfall sample than in the pre-landfall sample (as shown in fig. 5B for LA-26).

Table 18. Statistical comparisons of contaminant concentrations in pre-landfall samples to those in post-landfall samples from the Deepwater Horizon oil spill, Gulf of Mexico, 2010: organic contaminants in water.

[Significant p-values are shaded yellow (p<0.01) or orange (p<0.05). **Abbreviations**: A, no detections remained after censoring at optimal censoring threshold; B, no detections in paired dataset; C, no detections remain after blank censoring; mg/L, milligrams per liter; n, number of sample pairs; na, not applicable; nc, not censored at optimal censoring threshold because no detections remained after blank censoring; ns, not significant at 0.05 level in a 2-sided test; PPW, paired Prentice-Wilcoxon; µg/L, micrograms per liter; <, less than; ≤, less than or equal to; –, PPW test was not run]

| Analyte | Units | Optimal censoring threshold[1] | Paired Prentice–Wilcoxon test | | | |
			[2]n	p-value	Sampling period with significantly higher concentration	Reason no PPW test was run
Acenaphthene	µg/L	0.28	–	–	–	A
Acenaphthylene	µg/L	0.3	–	–	–	A
Anthracene	µg/L	0.39	–	–	–	A
Benzene	µg/L	0.34	44	0.0833	ns	na
Benzo[a]anthracene	µg/L	0.26	46.5	0.3173	ns	na
Benzo[a]pyrene	µg/L	0.33	46.5	0.3173	ns	na
Benzo[b]fluoranthene	µg/L	0.3	46.5	0.3173	ns	na
Benzo[k]fluoranthene	µg/L	0.4	–	–	–	B
Bis(2-chloroethyl)ether	µg/L	2.1	–	–	–	B
Bis(2-ethylhexyl) phthalate	µg/L	7.4	41.5	0.3173	ns	na
Carbon, organic	mg/L	3	40	0.0001	Post-landfall	na
Carbon disulfide	µg/L	0.5	41	0.3173	ns	na
Chloromethane	µg/L	0.53	–	–	–	A
Chrysene	µg/L	0.33	45	0.3173	ns	na
Dibenzo[a,h]anthracene	µg/L	0.42	–	–	–	B
Dichloromethane	µg/L	nc	–	–	–	C
Diethyl phthalate	µg/L	0.61	–	–	–	A
Di-*n*-butyl phthalate	µg/L	2	–	–	–	A
Ethylbenzene	µg/L	nc	–	–	–	C
Fluoranthene	µg/L	0.3	–	–	–	A
Fluorene	µg/L	0.33	–	–	–	A
Indeno(1,2,3-cd)pyrene	µg/L	0.38	–	–	–	B
Isophorone	µg/L	0.61	–	–	–	A
Naphthalene	µg/L	nc	–	–	–	C
Pentachlorophenol	µg/L	3.1	–	–	–	A
Phenanthrene	µg/L	0.32	–	–	–	A
Phenol	µg/L	1.5	–	–	–	A
Pyrene	µg/L	0.35	–	–	–	A
Toluene	µg/L	0.7	44	0.0144	Post-landfall	na
Tribromomethane	µg/L	0.58	–	–	–	A
Trichloromethane	µg/L	0.6	44	0.3173	ns	na
Xylenes, total	µg/L	1.6	44	0.3173	ns	na

[1]Lowest detection threshold that maximizes the number of quantifiable detections, minimizes the number of indeterminate samples, and has ≤5 to 8 percent indeterminate samples, depending on sample size.

[2]Non-integer indicates data missing for one member of a sample pair.

Toluene was not detected at greater than 0.7 µg/L in pre-landfall samples but was detected in six post-landfall samples, only five of which were in the paired data set and therefore appear in figure 5A. The significant PPW test result was influenced by the toluene detections in post-landfall samples from five sites: MS-37, FL-3, FL-4, FL-25, and FL-5 (fig. 5A). Additional BTEX compounds—benzene and xylenes—were detected in samples from two of these sites: MS-37 and FL-25.

The detection of BTEX compounds in post-landfall beach-water samples does not necessarily indicate the presence of M-1 oil. Weathered M-1 oil, which was collected on April 27, 2010, contained no detectable BTEX compounds; of the aliphatic and cyclic hydrocarbons detected, the lowest molecular-weight compound detected was the alkane n-C14 (State of Florida Oil Spill Academic Task Force, 2010), and BTEX compounds were not detected in surface-oil samples approaching the near shore environment (Atlas and Haven, 2011). BTEX compounds are volatile and tend to be rapidly removed from seawater by evaporation, and to a lesser extent by sorption to particles and sediment, biodegradation, and photolysis (Neff, 2002). Other sources of BTEX compounds to the GOM include produced water (Neff, 2002; Neff and others, 2011), deposition of airborne hydrocarbons from combustion sources, and natural oil and gas seeps (Continental Shelf Associates, 1997). Nevertheless, high concentrations of BTEX compounds, including up to 30 µg/L toluene, were reported in a plume trending southwest from the M-1 well at about 1,100 m depth in June 2010 (Reddy and others, 2012); it was concluded that although the ultimate fate of these compounds in the deep-water plume was unknown, the apportionments of hydrocarbon transfers to the water column and atmosphere appeared to be very different for a deep-water spill compared to a sea-surface oil spill. In the present study, water samples were collected at wadable depths near the shore, and toluene was detected in post-landfall water samples from six sites. Three of the six sites with toluene detections, MS-37, AL-7, and FL-3, were reported to have the M-1 oil fingerprint in corresponding post-landfall samples of sediment, tarballs, or both (Rosenbauer and others, 2010)—thus providing direct evidence of M-1 oil landfall at those sites at the time of post-landfall sampling—but the other three sites with toluene detections, FL-4, FL-25, and FL-5, did not show evidence of M-1 oil. No evidence of M-1 oil was found in 69 pre-landfall sediment samples analyzed by Rosenbauer and others (2011), although a tarball from one site, FL-18, was similar to M-1 oil, as discussed later in the report.

Comparison with Benchmarks for Human Health and Aquatic Life

Benchmark comparisons were made for all field samples, including primary environmental samples and field replicates. Benchmark exceedances for organics in water by individual sample are listed in appendix table 3-1, and the results are summarized in table 19. For those organic compounds with benchmarks, 253 water samples were analyzed: 196 pre-landfall samples from 70 sites and 57 post-landfall samples from 49 sites. Not every organic compound was analyzed in every sample, as indicated in appendix table 3-1. Of the 253 water samples, 138 samples were analyzed for PAHs and BTEX compounds, 86 samples for BTEX compounds only, and 29 samples for PAHs only.

Human-health benchmarks are available for 11 organic contaminants analyzed in water (table 5C). None of these benchmarks were exceeded by any water samples in this study.

Aquatic-life benchmarks used in the present study include the USEPA's toxic-unit benchmarks for mixtures of PAH and BTEX compounds (table 5A), as well as supplementary aquatic-life benchmarks for 72 individual organic contaminants (table 5B). One water sample exceeded the USEPA's chronic toxic-unit benchmark for PAH and BTEX compound mixtures (table 19, appendix 3-1). As noted previously, this benchmark assumes additive toxicity for compounds with the same mechanism of action, and a $\sum TU_i$ value greater than 1 indicates that chronic toxicity to aquatic life is likely. The single water sample exceeding this benchmark was the post-landfall sample from the Mississippi River at South Pass, Louisiana (site LA-35), for which the chronic $\sum TU_i$ value was 2.4. This is substantially higher than the corresponding chronic $\sum TU_i$ value of less than 10^{-4} for the pre-landfall sample collected at this site. Neither the post-landfall nor the pre-landfall sediment from site LA-35 contained the M-1 oil fingerprint (Rosenbauer and others, 2010 and 2011).

Of the 72 individual organic contaminants analyzed in this study that have aquatic-life benchmarks (table 5B), not all were analyzed in every water sample (see appendix table 3-1). However, none of the aquatic-life benchmarks for any individual organic contaminants were exceeded by any water samples in this study.

Of individual organic contaminants with benchmarks, recovery in matrix spikes was less than 70 percent for six contaminants—4-nitrophenol, benzo[a]pyrene, hexachlorobutadiene, hexachlorocyclopentadiene, hexachloroethane, and N-nitrosodiphenylamine—indicating that the measured concentration could be biased low.

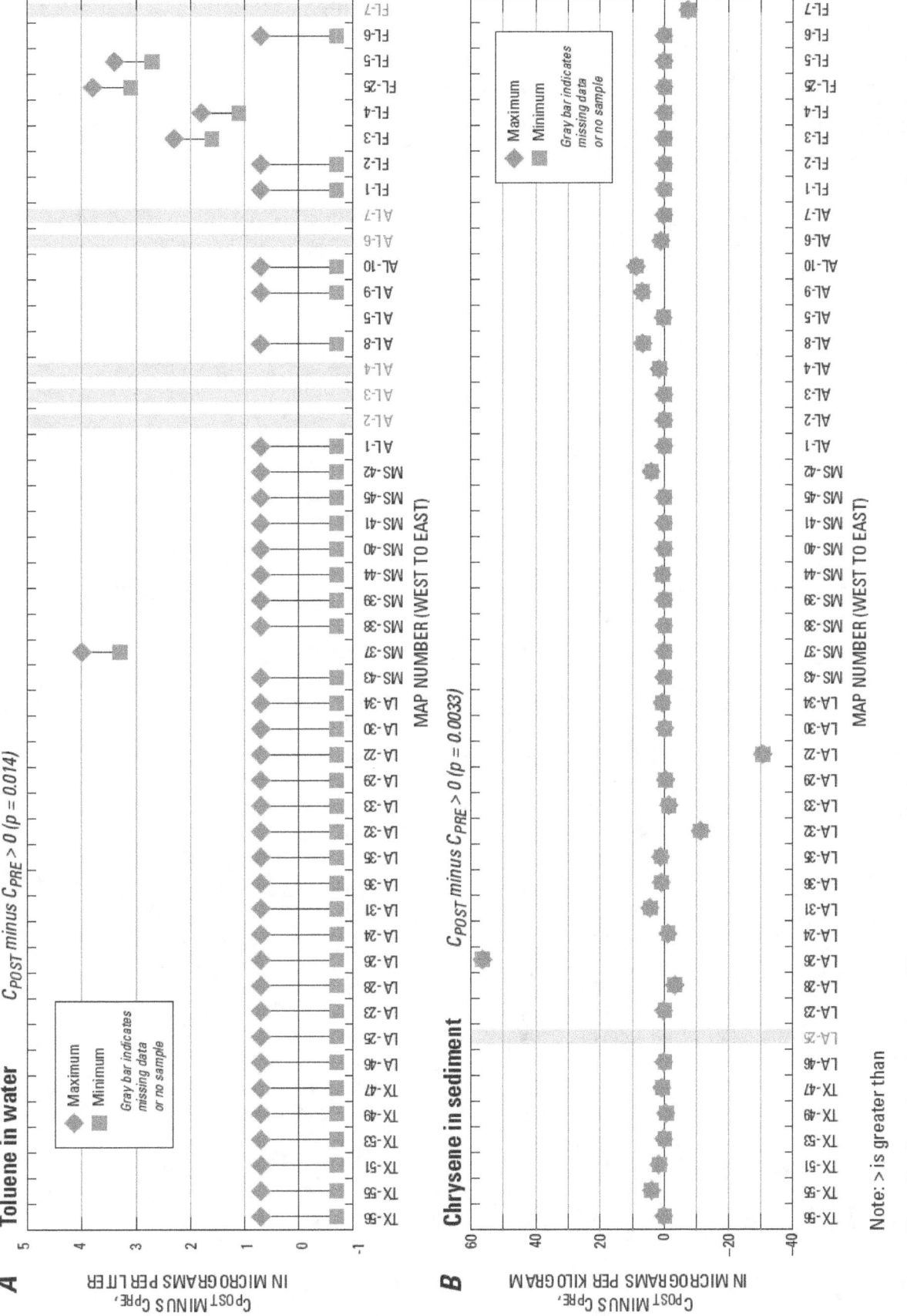

Figure 5. For selected analytes, difference between organic contaminant concentrations in samples collected during post-landfall and pre-landfall periods (C_{post} minus C_{pre}) by the USGS at Deepwater Horizon oil spill sampling sites along the Gulf of Mexico coast, 2010, from west to east, for (A) toluene in water, (B) chrysene in sediment, (C) C1-alkylated chrysenes in sediment, (D) C2-alkylated fluorenes in sediment, (E) C3-alkylated dibenzothiophenes in sediment, (F) C4-alkylated phenanthrenes/anthracenes in sediment, (G) naphthalene in sediment, and (H) oil and grease in sediment.

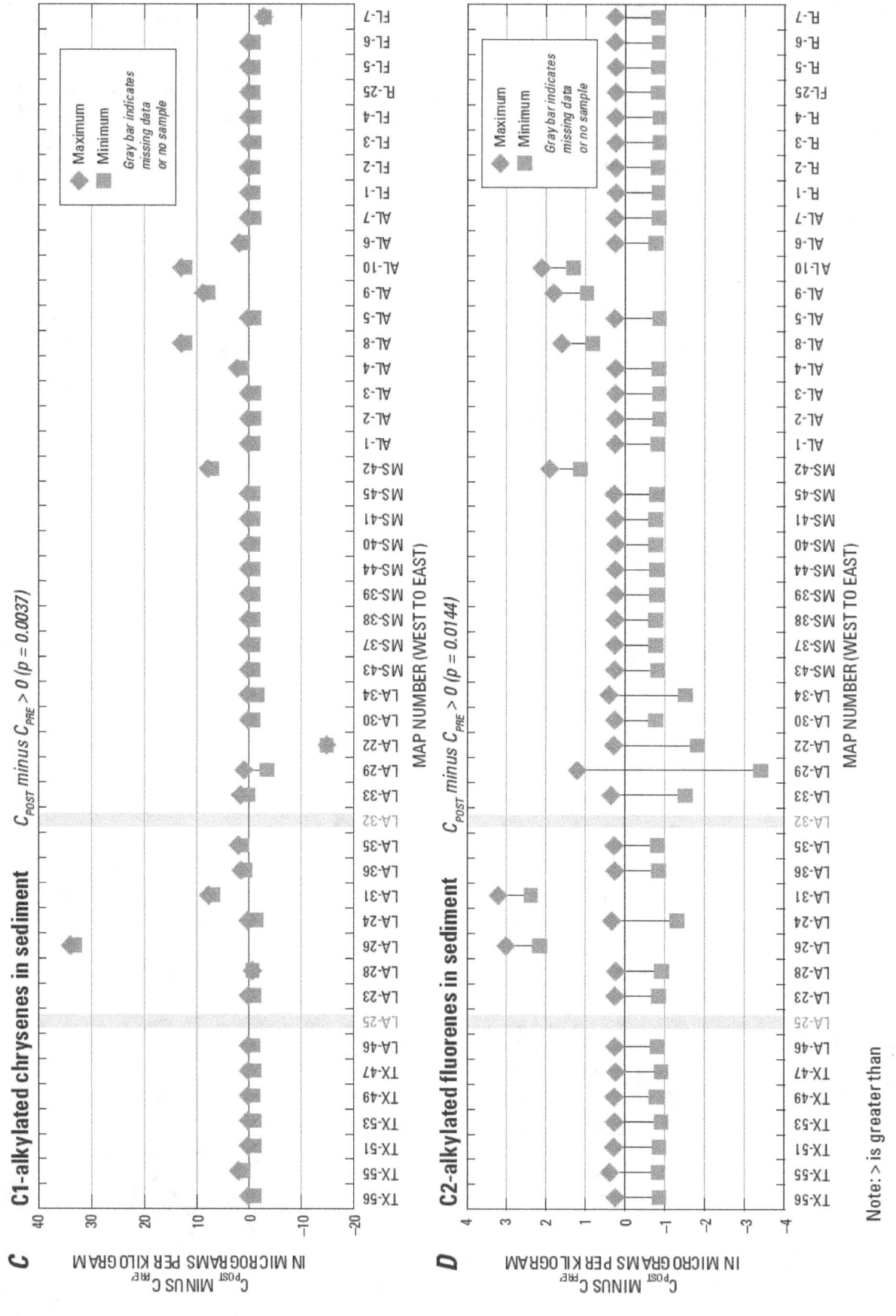

Note: > is greater than

Figure 5.—Continued

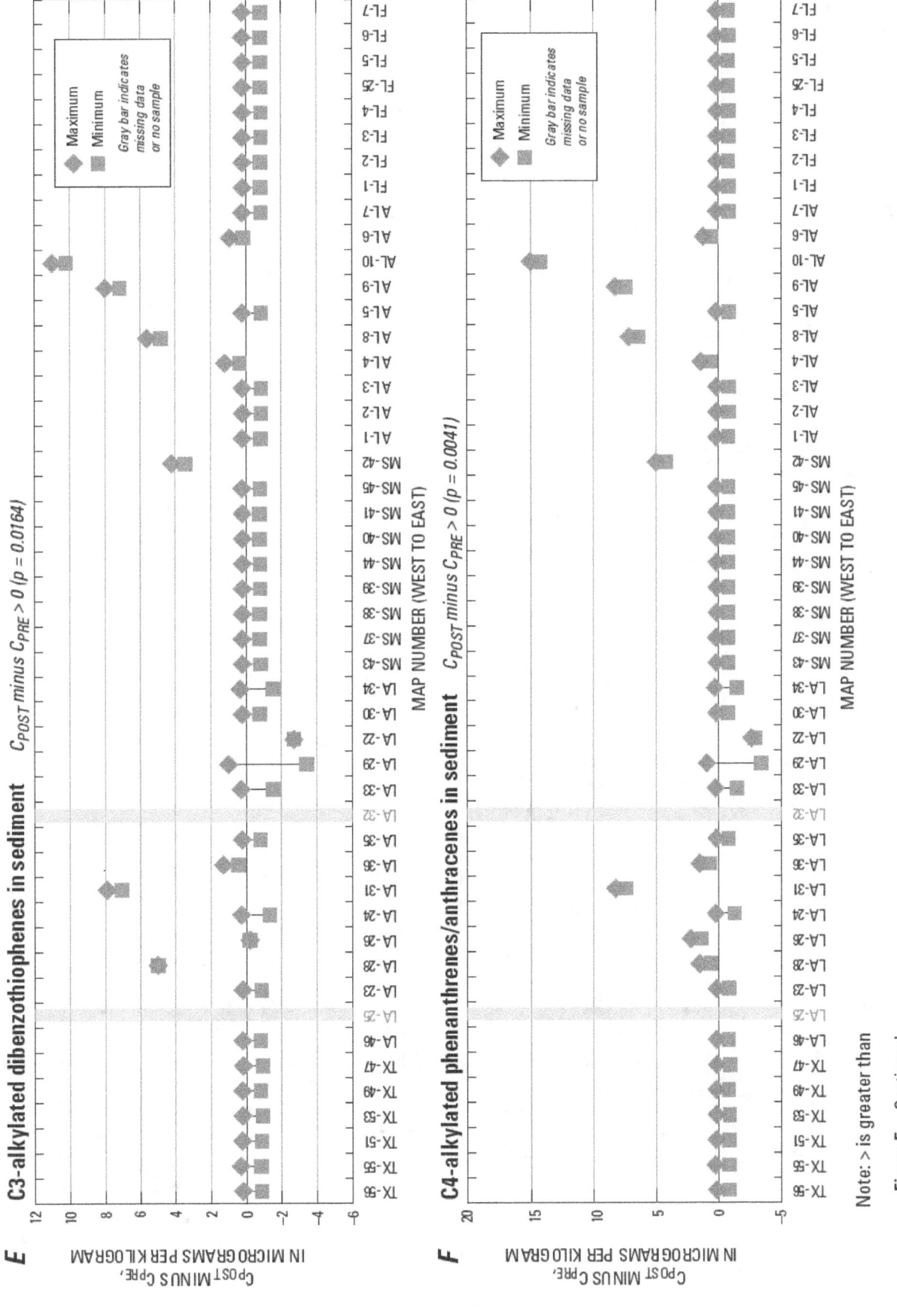

Note: > is greater than

Figure 5.—Continued

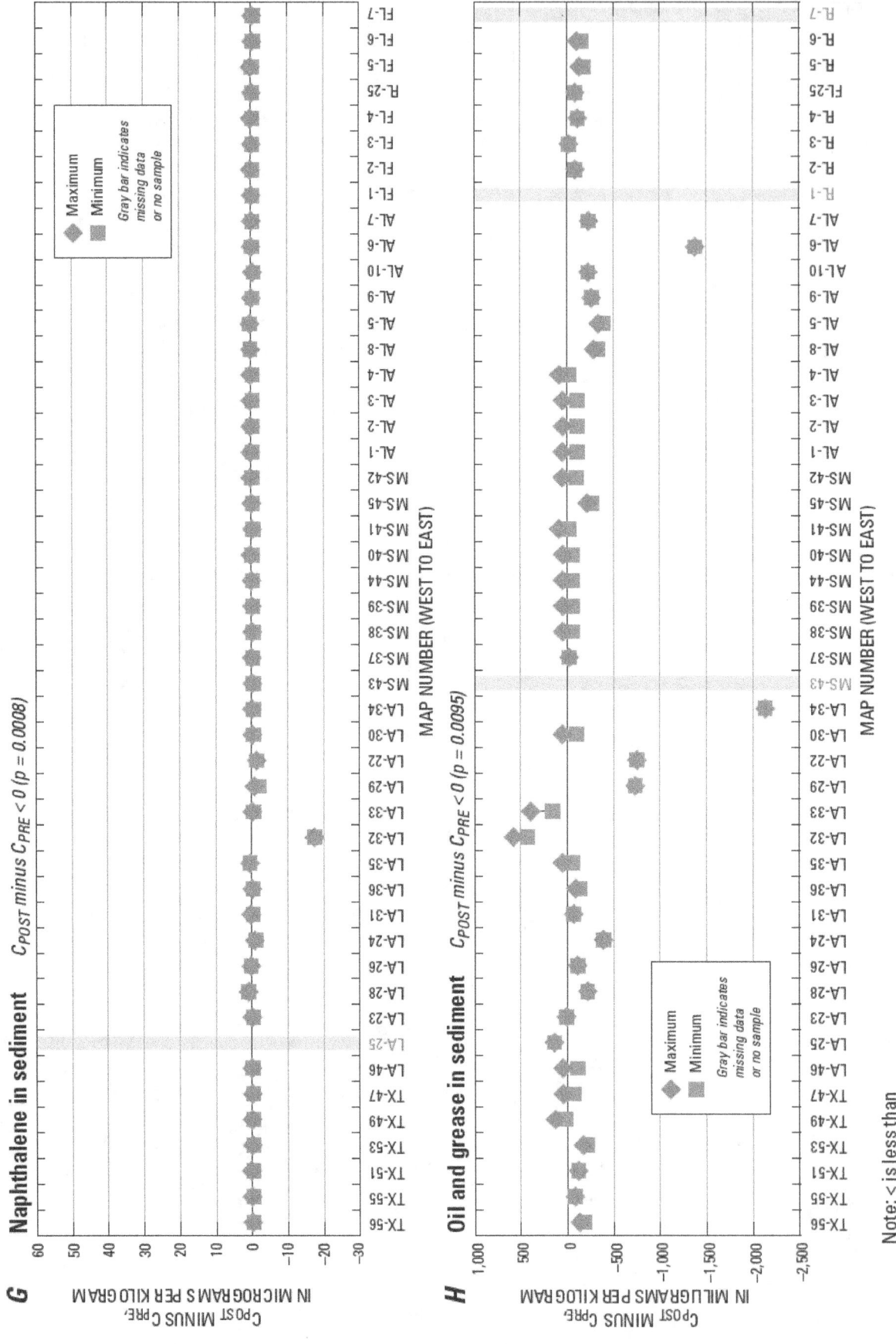

Note: < is less than

Figure 5.—Continued

Table 19. Summary of benchmark exceedances for organic contaminants in water and sediment from the Deepwater Horizon oil spill, Gulf of Mexico, 2010.

[Abbreviations: BTEX, benzene, toluene, ethylbenzene, xylene and related compounds; ESBTU, equilibrium-partitioning sediment benchmark toxic unit for PAH mixtures in sediment; MW, molecular weight; PAH, polycyclic aromatic hydrocarbon; post, post-landfall samples; pre, pre-landfall samples; SV, screening value; TU, toxic unit benchmark for PAH-BTEX mixtures in water; –, no benchmarks exceeded]

State	Water				Sediment					
	Number of sites	Total number of samples[1]	Number of samples with exceedances[2]	Benchmarks exceeded[2]	Number of sites	Total[3]	Number of samples			Upper SV and ESBTU benchmarks exceeded[7]
							In minimal effect range[4]	In possible effect range[5]	In probable effect range[6]	
Pre-landfall sampling period										
Texas	10	24	0	–	10	12	8	0	4	Total PAH, Low-MW PAH, chronic ESBTU
Louisiana	15	35	0	–	15	19	8	1	10	Total PAH, Low-MW PAH
Mississippi	9	25	0	–	9	11	11	0	0	–
Alabama	10	26	0	–	10	14	12	2	0	–
Florida	26	86	0	–	26	52	42	0	10	Total PAH, Low-MW PAH, High-MW PAH
Total pre-landfall	70	196	0	–	70	108	81	3	24	
Post-landfall sampling period										
Texas	6	9	0	–	6	10	5	1	4	Total PAH, Low-MW PAH
Louisiana	16	18	1	Chronic TU	15	17	5	0	12	Total PAH, Low-MW PAH, High-MW PAH
Mississippi	9	10	0	–	9	10	9	0	1	Total PAH
Alabama	10	11	0	–	10	11	8	0	3	Total PAH, Low-MW PAH, High-MW PAH
Florida	8	9	0	–	8	9	8	0	1	Total PAH
Total post-landfall	49	57	1		48	57	35	1	21	
Total (all samples)	119	253	1		118	165	116	4	45	

[1] The total number of water samples includes field replicate samples, and counts separately samples sent to different laboratories for analysis. About 70 percent of water samples were analyzed for PAHs and BTEX compounds, and 30 percent were analyzed for BTEX compounds at a different laboratory. Samples typically were collected from a given site on the same date.

[2] Organic contaminants in water were compared to aquatic-life benchmarks in tables 5A and 5B, and human-health benchmarks in table 5C.

[3] The total number of sediment samples includes field replicate samples. Samples were analyzed for PAHs, alkylated PAHs, and a few additional semivolatile organic compounds. Samples typically were collected from a given site on the same date.

[4] All constituent concentrations were below all applicable benchmarks (table 6C); there is no indication of adverse effects on benthic organisms within this range.

[5] One or more constituents exceeded a lower SV benchmark, but no elements exceeded an upper SV benchmark; adverse biological effects occasionally occur within this range.

[6] One or more constituents exceeded an upper SV benchmark; adverse biological effects frequently occur within this range.

[7] Organic contaminants in sediment were compared to sediment-quality benchmarks in tables 5D and 5E.

Benchmark results in appendix table 3-1 and summary statistics in table 17 are footnoted to indicate this. The single observed benchmark exceedance of the chronic TU benchmark for PAH and BTEX compounds by one post-landfall sample from site LA-35 must be considered in light of the QC data for organic contaminants in water. Of the compounds included in this benchmark, one BTEX compound (benzene) and one PAH compound (naphthalene) were detected in field or trip blanks associated with post-landfall samples; therefore, data for these two analytes were censored at five times the blank concentration to minimize the probability that incidental contamination contributed to the reported concentrations and any consequent benchmark exceedances. In the case of the LA-35 sample that exceeded the chronic TU benchmark, however, neither benzene nor naphthalene was detected in the sample; therefore, their concentrations were set to zero when computing the TU values for this sample, following the USEPA's calculation procedure and examples, which assume that censored values are equivalent to zero (Mount, 2010). Therefore, incidental contamination by benzene or naphthalene did not contribute to the chronic TU benchmark exceedance in the post-landfall sample at site LA-35.

Because there was only one benchmark exceedance, Fisher's exact test was not performed for organic contaminants in water. When chronic $\sum TU_i$ values for all 47 pairs of pre-landfall and post-landfall samples were compared, there was no significant difference between the two sampling periods (sign test, $p > 0.05$). In addition, acute $\sum TU_i$ values for PAH and BTEX compound mixtures were not greater than 1 in any water samples, and there was no significant difference in acute $\sum TU_i$ values between pre-landfall and post-landfall samples (sign test, $p > 0.05$). Again, these comparisons need to be qualified because reporting levels varied among analytes and between the two sampling periods, and concentrations were not censored to a single detection threshold prior to calculation of benchmark $\sum TU_i$ values, which were calculated following the standard USEPA procedure. Because reporting levels for many analytes were higher in post-landfall samples than in pre-landfall samples, setting nondetections equal to zero could underestimate benchmark exceedance rates in post-landfall samples relative to pre-landfall samples. Also, of the 47 sites with paired data, 6 pre-landfall sites were missing data for BTEX compounds, so the benchmark $\sum TU_i$ values for these pre-landfall samples were computed for PAHs only.

Organic Contaminants in Sediment

Most organic contaminants in sediment were determined by a single laboratory (TestAmerica Laboratory in Vermont) in samples from both sampling periods. These contaminants include parent PAHs and alkylated PAH groups, which are of potential concern from the oil spill (U.S. Environmental Protection Agency, 2011b). Fourteen additional organic contaminants, mostly individual alkylated PAH compounds, were analyzed only in pre-landfall samples by the USGS NWQL. Also, 44 miscellaneous SVOCs were analyzed in post-landfall but not pre-landfall samples; these include chlorinated phenols, nitroaromatic compounds, chlorinated alkanes and alkenes, nitroso compounds, and phthalate esters.

Reporting levels for organics in sediment varied somewhat for a given compound, but not as widely as for organics in water, and reporting levels were comparable for pre-landfall and post-landfall samples (appendix 2).

Contaminant Occurrence

The detection frequencies and percentile concentrations of organic contaminants in sediment are shown in table 20. Of the 14 organic contaminants analyzed only in pre-landfall samples, excluding TOC, 8 contaminants were detected in 1 to 4 samples each. Six of these were individual alkylated PAH compounds that also were included in determination of alkylated PAH groups (for example, 1,2-dimethylnaphthalene would be included in the C-2 naphthalenes group) by TestAmerica Laboratory in Vermont. The remaining two were 9,10-anthraquinone and the mixture of petroleum hydrocarbons.

There were 52 organic contaminants, plus organic carbon, analyzed in both pre-landfall and post-landfall samples: 19 parent PAHs, 5 individual alkylated PAHs, 22 alkylated PAH groups, 5 SVOCs, and oil and grease. Of the 52 analytes, 49 were detected in pre-landfall samples and 50 in post-landfall samples, with 47 analytes detected in samples from both sampling periods. Only two analytes were not detected in any samples: the SVOCs, hexachlorobenzene and diethyl phthalate. PAH detection frequencies above the optimal censoring threshold for each analyte ranged from 3 to 64 percent for parent PAHs and 0 to 33 percent for alkylated PAH groups; because of variable reporting limits, there were some indeterminate samples (table 20). Figure 4C shows an example of the concentration distribution observed in sediment samples for the alkylated PAH group, C3-alkylated fluorenes. The reporting levels for C3-alkylated fluorene tend to be lower for post-landfall than pre-landfall samples, which means that uncensored detection frequencies will not provide a fair comparison of occurrence in the two sampling periods. After censoring at an optimal threshold of 1.8 μg/kg, the detection frequency for C3-alkylated fluorenes was higher in post-landfall samples at 15 percent than in pre-landfall samples, where it was 1 percent.

Table 20. Summary statistics for organic contaminants in sediment from the Deepwater Horizon oil spill, Gulf of Mexico, 2010.

This table is presented as a Microsoft© Excel spreadsheet. It can be accessed and downloaded at URL http://pubs.usgs.gov/sir/2012/5228.

Comparison of detection frequencies among contaminants with different reporting levels should be done at a common detection threshold (table 20). For example, the parent PAH, chrysene, was detected above its optimal censoring threshold of 0.23 µg/kg in 50 percent of post-landfall samples, compared to 33, 29, 21, and 13 percent of post-landfall samples for the C-1, C-2, C-3, and C-4 alkylated chrysenes detected above their respective optimal censoring thresholds, which are 1.5, 1, 1, and 1.3 µg/kg. When a common detection threshold of 1.5 µg/kg was applied, the 29-percent detection frequency for chrysene was then comparable to detection frequencies for the C-1 and C-2 alkylated chrysenes of 33 and 27 percent, respectively, and it was closer to those for C-3 and C-4 alkylated chrysenes, which were 19 percent and 10 percent, respectively.

Thirteen PAHs—four parent and nine alkylated—were detected at or above concentrations of 1.5 µg/kg in more than 20 percent of post-landfall samples, whereas four parent PAHs were detected at or above the 1.5 µg/kg threshold in more than 20 percent of pre-landfall samples (table 20). Overall, PAH detection frequencies in sediment (table 20) tended to be higher than in water samples (table 17), which is expected because PAHs are hydrophobic and tend to sorb to organic material.

Comparison of Pre-Landfall to Post-Landfall Samples

Of 54 organic contaminants analyzed during both pre-landfall and post-landfall sampling periods, and for at least 80 whole-sediment samples, there were enough quantified detections above the optimal censoring threshold to make a statistical comparison of pre-landfall to post-landfall samples for 49 contaminants (table 21). Parent PAHs accounted for 19 of these contaminants, and alkylated PAHs accounted for 26 contaminants. Of these 49 contaminants, 22 showed a significant difference between pre-landfall and post-landfall samples in PPW tests ($p<0.05$; table 21). Concentrations were significantly higher in post-landfall samples for 20 contaminants, which included 3 PAHs and 17 alkylated PAH groups, and in pre-landfall samples for two contaminants, naphthalene and oil and grease. The difference between concentrations in post-landfall and pre-landfall sediment samples ($C_{post} - C_{pre}$) at individual sites along the GOM coast, from west to east, is shown in figures 5B to 5H for some example contaminants with significantly higher concentrations during one sampling period than the other. The examples in figures 5B–5F are PAHs that had significantly higher concentrations in post-landfall than pre-landfall samples; they represent various PAH ring structures and various degrees of alkylation. Figures 5G and 5H show naphthalene and oil and grease, respectively, for which concentrations were significantly higher in pre-landfall samples than post-landfall samples. For the three parent PAHs with significant PPW tests, chrysene, benzo[b]fluoranthene, and benzo[e]pyrene, there were high post-landfall sample concentrations at one site— LA-26 (for example, see chrysene in fig. 5B). About 1.5 to 2 times more sample pairs had a positive difference when subtracting pre-landfall samples from post-landfall samples than had a negative difference for these three PAHs, which is consistent with the significant test result.

In contrast, the significant results for 15 of 17 alkylated PAHs reflected particularly high concentrations in post-landfall samples at seven sites: LA-28, LA-26, LA-31, MS-42, AL-8, AL-9, and AL-10 (for example, see figs. 5C–5F). Five of these sites showed post-landfall evidence of M-1 oil in sediment, tarballs, or both, on the basis of PAH fingerprinting by Rosenbauer and others (2010): Grand Isle Beach at State Park, Louisiana (LA-31); Petit Bois Island Beach, Mississippi (MS-42); and BLM-1 (AL-8), BLM-2 (AL-9), and Fort Morgan BLM-3 (AL-10) in Alabama. Notably, 16 of the 17 alkylated PAHs with significantly higher concentrations in post-landfall samples were identified as relatively abundant components of weathered M-1 oil (State of Florida Oil Spill Academic Task Force, 2010). Chrysene and alkylated PAHs, however, are characteristic of petrogenic PAHs (those originating from petroleum and petroleum products) in general (Iqbal and others, 2008).

For two sites, LA-28 and LA-26, with large positive differences in alkylated PAHs when pre-landfall concentrations were subtracted from post-landfall concentrations, there was no evidence of the M-1 oil fingerprint in the post-landfall sediment samples (Rosenbauer and others, 2010). The most abundant PAH compounds in these samples were consistent with pyrogenic sources, which result from combustion of organic matter and fossil fuels. In the LA-28 sample, the most abundant PAH compound was anthracene, which is produced during rapid, high temperature pyrosynthesis but does not persist during the slow diagenesis leading to the generation of fossil fuels (Iqbal and others, 2008). In the LA-26 sample, the most abundant PAHs were fluoranthene and pyrene, and alkylated PAH concentrations were generally lower than the corresponding parent PAHs, which are characteristic of pyrogenic sources of PAHs.

Table 21. Statistical comparisons of contaminant concentrations in pre-landfall samples and post-landfall samples from the Deepwater Horizon oil spill, Gulf of Mexico, 2010: organic contaminants in sediment.

[Significant p-values are shaded yellow (p<0.01) or orange (p<0.05). Abbreviations: g/kg, gram per kilogram; mg/kg, milligram per kilogram; n, number of pre-landfall and post-landfall sample pairs in PPW test; na, OC-normalization is not applicable; nd, optimal threshold not determined; ns, not significant at 0.05 level in a 2-sided test; OC, organic carbon; Post, post-landfall period; PPW, paired Prentice-Wilcoxon; Pre, pre-landfall period; µg/kg, microgram per kilogram; µg/g-oc, microgram per gram of sediment organic carbon; —, PPW test was not run; <, less than]

| Analyte | Dry-weight concentrations | | | | | OC-normalized concentrations[3] | | | | |
| | Units | Optimal censoring threshold[1] | Paired Prentice-Wilcoxon test | | | Units | Optimal censoring threshold[1] | Paired Prentice-Wilcoxon test | | |
			[2]n	p-value	Sampling period with significantly higher concentration			[2]n	p-value	Sampling period with significantly higher concentration
1-Methylnaphthalene	µg/kg	0.4	44	0.6054	ns	µg/g-oc	0.54	—[4]	—	—
1-Methylphenanthrene	µg/kg	0.38	45.5	0.0975	ns	µg/g-oc	0.44	40	0.0696	ns
26-Dimethylnaphthalene	µg/kg	0.47	45.5	0.5378	ns	µg/g-oc	0.56	43.5	0.1573	ns
2-Methylnaphthalene	µg/kg	0.53	45	0.9755	ns	µg/g-oc	0.64	—[4]	—	—
Acenaphthene	µg/kg	0.36	44.5	0.5983	ns	µg/g-oc	0.48	45	0.9875	ns
Acenaphthylene	µg/kg	0.38	45.5	0.9505	ns	µg/g-oc	0.46	45.5	0.1573	ns
Anthracene	µg/kg	0.38	45.5	0.6515	ns	µg/g-oc	0.46	45.5	0.581	ns
Benzo[a]anthracene	µg/kg	0.25	42.5	0.7539	ns	µg/g-oc	0.48	47	0.2561	ns
Benzo[a]pyrene	µg/kg	0.25	42.5	0.5311	ns	µg/g-oc	0.48	47	0.111	ns
Benzo[b]fluoranthene	µg/kg	0.23	47	0.0333	Post	µg/g-oc	0.44	43	0.0512	ns
Benzo[e]pyrene	µg/kg	0.38	45	0.0219	Post	µg/g-oc	0.72	41.5	0.0074	Post
Benzo[ghi]perylene	µg/kg	0.25	43.5	0.2264	ns	µg/g-oc	0.5	45	0.1742	ns
Benzo[k]fluoranthene	µg/kg	0.4	45.5	0.3122	ns	µg/g-oc	0.76	43	0.1257	ns
Biphenyl	µg/kg	0.47	45	0.3173	ns	µg/g-oc	0.56	—[4]	—	—
Bis(2-ethylhexyl) phthalate	µg/kg	80	44	0.5816	ns	µg/g-oc	110	44.5	0.3173	ns
C1 Chrysene	µg/kg	1.5	46	0.0037	Post	µg/g-oc	1.7	45	0.0016	Post
C1 Dibenzothiophenes	µg/kg	1	44.5	0.1727	ns	µg/g-oc	1.7	45	0.1932	ns
C1 Fluoranthenes/pyrenes	µg/kg	0.91	44	0.0187	Post	µg/g-oc	1.7	45	0.0297	Post
C1 Fluorenes	µg/kg	1.5	45.5	0.1573	ns	µg/g-oc	1.7	—[4]	—	—
C1 Naphthalenes	µg/kg	1.3	44.5	0.1573	ns	µg/g-oc	1.7	—[4]	—	—
C1 Phenanthrenes/anthracenes	µg/kg	1.5	46	0.0224	Post	µg/g-oc	1.7	45	0.0082	Post
C2 Chrysenes	µg/kg	1	44.5	0.0123	Post	µg/g-oc	1.7	45	0.0027	Post
C2 Dibenzothiophenes	µg/kg	1	44.5	0.0322	Post	µg/g-oc	1.7	45	0.0191	Post
C2 Fluoranthenes/pyrenes	µg/kg	1.3	45	0.0379	Post	µg/g-oc	1.7	45	0.0101	Post
C2 fluorenes	µg/kg	1.3	44.5	0.0144	Post	µg/g-oc	1.7	45	0.0254	Post
C2 naphthalenes	µg/kg	0.91	44.5	0.6312	ns	µg/g-oc	1.7	45	0.9875	ns
C2 Phenanthrenes/anthracenes	µg/kg	1	44.5	0.0091	Post	µg/g-oc	1.7	45	0.0027	Post
C3 Chrysenes	µg/kg	1	44.5	0.0136	Post	µg/g-oc	1.7	45	0.0144	Post
C3 Dibenzothiophenes	µg/kg	1	44.5	0.0164	Post	µg/g-oc	1.7	45	0.0176	Post
C3 Fluoranthenes/Pyrenes	µg/kg	1.5	46	0.0358	Post	µg/g-oc	1.7	45	0.0144	Post
C3 Fluorenes	µg/kg	1.8	46	0.0144	Post	µg/g-oc	1.7	45	0.0254	Post
C3 Naphthalenes	µg/kg	0.91	44.5	0.6595	ns	µg/g-oc	1.7	45	0.9875	ns
C3 Phenanthrenes/anthracenes	µg/kg	1	44.5	0.0024	Post	µg/g-oc	1.7	45	0.0027	Post

Table 21. Statistical comparisons of contaminant concentrations in pre-landfall samples and post-landfall samples from the Deepwater Horizon oil spill, Gulf of Mexico, 2010: organic contaminants in sediment.—Continued

[Significant p-values are shaded yellow (p<0.01) or orange (p<0.05). **Abbreviations:** g/kg, gram per kilogram; mg/kg, milligram per kilogram; n, number of pre-landfall and post-landfall sample pairs in PPW test; na, OC-normalization is not applicable; nd, optimal threshold not determined; ns, not significant at 0.05 level in a 2-sided test; OC, organic carbon; Post, post-landfall period; PPW, paired Prentice-Wilcoxon; Pre, pre-landfall period; µg/kg, microgram per kilogram; µg/g-oc, microgram per gram of sediment organic carbon; —, PPW test was not run; <, less than]

| | Dry-weight concentrations | | | | | OC-normalized concentrations[3] | | | | |
| | Paired Prentice-Wilcoxon test | | | | | | Paired Prentice-Wilcoxon test | | | |
Analyte	Units	Optimal censoring threshold[1]	[2]n	p-value	Sampling period with significantly higher concentration	Units	Optimal censoring threshold[1]	[2]n	p-value	Sampling period with significantly higher concentration
C4 Chrysenes	µg/kg	1.3	44.5	0.0144	Post	µg/g-oc	1.7	45	0.0254	Post
C4 Dibenzothiophenes	µg/kg	1.8	46	0.0047	Post	µg/g-oc	1.7	45	0.0082	Post
C4 Naphthalenes	µg/kg	1.2	44.5	0.0253	Post	µg/g-oc	1.7	45	0.0254	Post
C4 Phenanthrenes/anthracenes	µg/kg	1.3	45	0.0041	Post	µg/g-oc	1.7	45	0.0047	Post
Chrysene	µg/kg	0.23	47	0.0033	Post	µg/g-oc	0.44	43	0.0049	Post
Dibenzo[ah]anthracene	µg/kg	0.38	45	0.7296	ns	µg/g-oc	0.46	45	0.9875	ns
Dibenzothiophene	µg/kg	0.38	45.5	0.1491	ns	µg/g-oc	0.46	45.5	0.3173	ns
Fluoranthene	µg/kg	0.26	47	0.649	ns	µg/g-oc	0.5	45	0.7769	ns
Fluorene	µg/kg	0.44	45	1	ns	µg/g-oc	0.54	45	0.3173	ns
Indeno[123cd]pyrene	µg/kg	0.23	45.5	0.5065	ns	µg/g-oc	0.46	47	0.3389	ns
Naphthalene	µg/kg	0.29	40	0.0008	Pre	µg/g-oc	0.56	45.5	0.0044	Pre
Oil and grease	mg/kg	110	45.5	0.0095	Pre	µg/g-oc	nd[5]	—[5]	—	—
Organic carbon	percent	0.1	44	0.2478	ns	µg/g-oc	na	—	—	—
Perylene	µg/kg	0.22	41.5	0.6748	ns	µg/g-oc	0.44	43	0.9221	ns
Phenanthrene	µg/kg	0.23	44	0.3913	ns	µg/g-oc	0.44	43.5	0.9247	ns
Pyrene	µg/kg	0.24	45.5	0.1712	ns	µg/g-oc	0.48	47	0.2265	ns

[1]Lowest detection threshold at which the percentage of indeterminate samples is less than or equal to 5 percent.

[2]Non-integer indicates data missing for one member of a sample pair.

[3]Normalized contaminant concentrations (in µg/g-oc) were calculated by dividing the censored contaminant concentration (µg/kg) by the sediment organic-carbon concentration (g/kg).

[4]No detections remained after censoring at optimal censoring threshold.

[5]Organic-carbon data is not available for a sufficient number of pre-landfall samples to determine an optimal censoring threshold or run PPW test on organic carbon-normalized data for this constituent.

Naphthalene and oil and grease concentrations were significantly higher in pre-landfall sediment samples than post-landfall samples (table 21). Twenty-seven sample pairs have negative difference values when pre-landfall concentrations were subtracted from post-landfall concentrations, compared to only 5 pairs with positive difference values. Moreover, one site, LA-32, has a very large naphthalene difference value (fig. 5G). Similarly, oil and grease concentrations at 26 sites along the GOM coast had negative difference values when pre-landfall concentrations were subtracted from post-landfall concentrations, compared to 5 sites with positive difference values; concentrations in pre-landfill samples were substantially higher for sites LA-29, LA-22, LA-34, and AL-6 (fig. 5H). Oil and grease are operationally defined as hexane-extractable material, which includes relatively nonvolatile hydrocarbons, vegetable oils, animal fats, waxes, soaps, and greases (U.S. Environmental Protection Agency, 1998).

Because hydrophobic contaminants such as PAHs tend to be associated with organic carbon, it is possible that differences in the amount of organic carbon in pre-landfall and post-landfall samples could have caused or contributed to the significant differences in PAH concentrations. Therefore, the PPW tests were repeated after normalizing organic contaminant concentrations to the sediment-TOC content (table 21). Of the 20 PAHs with significantly higher concentrations in post-landfall samples, 19 continued to show a significant difference after organic-carbon normalization; the 20th had a p-value of 0.051, which is only slightly greater than the significance criterion of p<0.05. Sediment-TOC data were insufficient to normalize oil and grease concentrations; however, naphthalene concentrations were significantly higher in pre-landfall than in post-landfall samples even after organic-carbon normalization. Moreover, there was no significant difference in sediment-TOC content between the two sampling periods (table 21). These PPW test results indicate that the significant differences are not likely due to differing amounts of sediment-TOC in samples from pre-landfall and post-landfall periods.

The results of the present study, combined with direct evidence from the oil fingerprinting study by Rosenbauer and others (2010), indicate that M-1 oil could have contributed to the higher alkylated PAH concentrations measured at five sites, LA-31, MS-42, AL-8, AL-9, and AL-10, sampled in October 2010, relative to pre-landfall concentrations; however, other PAH sources, including other sources of oil, cannot

be excluded. There are many possible sources of oil-related contaminants in the GOM, including natural oil seepage, which is estimated at about one million barrels of petroleum hydrocarbons each year; various oil spills from production operations, which contribute approximately 74,000 barrels each year; transportation accidents; and unburned engine fuel (Operational Science Advisory Team, 2010). A previous study of PAH sources along the Louisiana coast (Iqbal and others, 2008) reported that approximately 50 percent of PAHs were from petrogenic sources; 36 percent were from pyrogenic sources; and 14 percent were from diagenetic sources, that is, the chemical or biological transformation of natural organic matter.

Comparison with Benchmarks for Aquatic Life

The USEPA ESBTU benchmarks address the additive toxicity of PAH and BTEX compound mixtures in sediment (table 5D). As noted previously, $\Sigma ESBTU_i$ values were calculated only for PAHs because BTEX compounds were not determined in sediment. One sediment sample exceeded the chronic ESBTU benchmark for PAH mixtures: the pre-landfall sample from Trinity Bay near Beach City, Texas (site TX-52). This site was outside the area of expected oil landfall and was not sampled during the post-landfall period. Notably, sediment-TOC concentrations in the present study were very low, having a median of 0.1 percent, which could affect bioavailability and potential toxicity. As previously noted, equilibrium-partitioning theory predicts PAH toxicity in sediments that have a TOC content of 0.2 percent or above (U.S. Environmental Protection Agency, 2002).

Empirical screening values (table 5E) for 20 individual PAHs, 3 PAH mixtures, and 24 other SVOCs in sediment were used to classify sites into one of three categories: the minimal-effect, possible-effect, and probable-effect ranges. Of 165 sediment samples analyzed for organic contaminants that have benchmarks, 116 samples (70 percent) had no lower or upper screening values exceeded by any of the organic contaminants determined in the sample, so these were in the minimal-effect range where no adverse effects would be expected; 45 samples (27 percent) exceeded one or more upper screening values and so were in the probable-effect range, where there is a high probability of adverse effects on aquatic life; and only 4 samples (2 percent) were in the possible-effect range (table 19; appendix table 3-2).

Twenty one out of 57 post-landfall samples (37 percent) exceeded one or more upper screening values compared to 24 out of 108 pre-landfall samples (22 percent). The reverse pattern holds for samples where no screening values were exceeded, so that no adverse effects are expected, which applied to 81 of 108 of pre-landfall samples (75 percent) and 35 of 57 post-landfall samples (61 percent). The only upper screening-value benchmarks exceeded were for PAH mixtures. Lower screening values were exceeded by PAH mixtures, a few individual PAHs, and occasionally by bis(2-ethylhexyl)phthalate. Although three PAH compounds (benzo[g,h,i]perylene, indeno[1,2,3-cd]pyrene, and perylene) were detected in laboratory reagent blanks associated with two post-landfall samples, it is unlikely that any incidental contamination contributed to benchmark exceedances for these two samples. Neither of these two samples exceeded the ESBTU for total PAHs, and only one empirical benchmark—a lower screening value for perylene—was exceeded by one of these samples. On the other hand, four organic contaminants with benchmarks had less than 50 percent recovery from matrix spikes, so their concentrations and contribution to benchmark exceedance could be biased low. These were acenaphthylene and naphthalene, which are PAH compounds included in the ESBTU; 1,2,4-trichlorobenzene, which had no benchmark exceedances; and N-nitrosodiphenylamine, which was not evaluated because the benchmark was below the reporting level.

For the five sites identified as having possible contributions to alkylated PAH concentrations from M-1 oil—LA-31, MS-42, AL-8, AL-9, and AL-10—PAH concentrations did not exceed ESBTU benchmarks. Chronic $\sum ESBTU_i$ values in post-landfall samples from these sites ranged from 0.17 to 0.29 and so were below the hazard index of 1; this indicates that PAH levels in these post-landfall samples were not high enough to cause toxicity to benthic organisms according to these criteria. On the other hand, these samples did exceed empirical upper screening-value benchmarks for total PAHs, indicating a high probability of toxicity to benthic organisms at these sites as indicated by other field studies (MacDonald and others, 2000; Ingersoll and others, 2001).

Because of differences in how various benchmarks are derived, it is not surprising that empirical benchmarks were exceeded more often than the ESBTU benchmarks. The empirical, upper screening values are probabilistic—they are associated with frequent occurrence of toxicity in field sediments, which often contain mixtures of contaminants. Exceedance of an empirical benchmark is an indicator that toxicity is likely; it does not guarantee toxicity, and concentrations above the benchmark do not necessarily cause

toxicity. In contrast, the ESBTU benchmark is causally based and designates concentrations expected to result in PAH-induced toxicity to benthic organisms.

Direct comparison between benchmark-exceedance frequencies for pre-landfall and post-landfall sampling periods must be qualified because, as noted previously, data from the two sampling periods do not represent exactly the same sites: 22 pre-landfall sites in Florida and Texas and 1 post-landfall site in Louisiana were only sampled during one sampling period (table 1). Also, 20 of the 71 total sites were sampled more than once during one or both sampling periods. Differences in benchmark exceedances, however, were evaluated for the paired-sample dataset, which excludes exceedance data for field replicate samples and for sites sampled during only one period. Fisher's exact test indicated there was no significant difference in the benchmark-exceedance frequency between pre-landfall and post-landfall samples in this dataset ($p > 0.05$). This was true for exceedance of both upper and lower screening-value benchmarks. When chronic $\sum ESBTU_i$ values for paired pre-landfall and post-landfall samples were compared, there was no significant difference between the two sampling periods (sign test, $p > 0.05$).

Trace and Major Elements and Nutrients in Water

For trace and major elements and nutrients in water, the USGS NWQL analyzed pre-landfall samples, and TestAmerica Laboratory in Florida analyzed post-landfall samples. For some trace elements and nutrients, the method used to analyze pre-landfall samples by the USGS NWQL was more sensitive than the method used for post-landfall samples by TestAmerica Laboratory in Florida.

Constituent Occurrence

The detection frequencies and percentile concentrations for trace and major elements and nutrients in beach water samples are shown in table 22. Detection frequencies are provided for a series of detection thresholds because a common detection threshold must be applied when comparing detection frequencies between pre-landfall and post-landfall samples or for two different constituents. Reporting levels for trace elements in water were highly variable because 77 percent of water samples were diluted prior to trace element analysis, at least in part because of high specific conductance values.

Table 22. Summary statistics for trace and major elements and nutrients in water from the Deepwater Horizon oil spill, Gulf of Mexico, 2010.

This table is presented as a Microsoft© Excel spreadsheet. It can be accessed and downloaded at URL http://pubs.usgs.gov/sir/2012/5228.

Several patterns of trace-element occurrence were observed. Uncensored detection frequencies for many constituents tended to be higher in pre-landfall samples than in post-landfall samples. For some constituents, however, this simply reflects the lower reporting levels used to analyze these constituents in pre-landfall samples. When data were censored to a common reporting level, detection frequencies and concentrations were similar (table 22). For example, zinc concentrations detected in pre-landfall water samples (filled circles in fig. 4D) were generally below the laboratory reporting levels for post-landfall samples (unfilled squares in fig. 4D). However, after censoring at the optimal censoring threshold of 80 µg/L, the detection frequencies for zinc in water samples from the two sampling periods were the same at about 2 percent (table 22). Additional examples of this pattern were found with lead, which was detected above a threshold of 20 µg/L in 2 to 3 percent of samples from both sampling periods, and iron, which was detected above a threshold of 500 µg/L in 41 to 42 percent of samples from both periods. Molybdenum in water (fig. 4E) showed a different pattern, in which uncensored detection frequencies were higher in pre-landfall samples than post-landfall samples, but the concentration distribution was higher in post-landfall samples. After censoring to the optimal threshold of 20 µg/L, the molybdenum detection frequency was actually higher in post-landfall samples, at 8 percent, than in pre-landfall samples, where it was 0 percent. Aluminum and manganese also showed greater detection frequencies above their respective optimal censoring thresholds in post-landfall than pre-landfall samples. The nutrients, phosphorus and ammonia, were more frequently detected in pre-landfall than post-landfall samples, even after censoring to a common detection threshold. Phosphorus concentrations (fig. 4F) in post-landfall water samples had to be blank-censored first to minimize the possibility that the detected concentrations were the result of incidental contamination. Because the blank-censoring procedure is intentionally conservative, this could have overestimated the extent of incidental contamination and thus lowered the post-landfall sample detection frequency. Similarly, ammonia plus organic nitrogen (fig. 4G) was blank-censored in post-landfall water samples because of detection in each of four field blanks for the post-landfall sampling period. The conservative blank-censoring procedure resulted in censored data with high reporting levels for all post-landfall samples. When detection frequencies were computed at the optimal censoring threshold of 2.4 mg/L as

nitrogen (N), the ammonia detection frequencies were zero in both sampling periods. Barium, calcium, magnesium, potassium, and sodium were detected in 100 percent of both pre-landfall and post-landfall water samples, although some concentrations in post-landfall samples were higher than in pre-landfall samples (for example, potassium in fig. 4H; appendix 2–3).

Comparison of Pre-Landfall to Post-Landfall Samples

Statistical comparisons of trace and major element and nutrient concentrations in water were made for 17 of the 26 constituents determined in water during both study periods (table 23). The other nine constituents had no detections remaining after censoring, so no comparisons were made. The PPW test indicated significant differences between concentrations in pre-landfall and post-landfall water samples for nine constituents: six trace or major elements had higher concentrations in post-landfall samples, and three nutrients had higher concentrations in pre-landfall samples (table 23). Concentrations were higher in post-landfall samples for barium, calcium, magnesium, molybdenum, potassium and sodium. These are all elements in seawater (Turekian, 1968), and barium sulfate is a standard additive in drilling mud (Argonne National Laboratory and others, 2012). By using molybdenum as an example, figure 6A shows the difference in molybdenum concentrations in water between post-landfall and pre-landfall samples at individual sites along the GOM coast from west to east. Many sites had censored data for one or both samples; these are represented by bars that touch or cross the x-axis, that is, where y equals zero. Eighteen sites showed a positive difference when pre-landfall samples were subtracted from post-landfall samples, indicating higher post-landfall sample concentrations than pre-landfall, and six sites showed a negative difference, indicating the opposite.

Three nutrients, ammonia as N, ammonia as NH_4, and phosphorus, showed statistically significant differences (table 23), having higher detection frequencies (table 22) and higher concentrations in pre-landfall samples than post-landfall samples (for example, fig. 6B). Data were insufficient to assess ammonia plus organic nitrogen. Statistical comparisons were not significant for organic nitrogen and dissolved nitrogen (table 23).

Table 23. Statistical comparisons of contaminant concentrations in pre-landfall samples to those in post-landfall samples from the Deepwater Horizon oil spill, Gulf of Mexico, 2010: trace and major elements and nutrients in water.

[Significant p-values are shaded yellow (p<0.01) or orange (p<0.05). **Abbreviations**: A, No quantified detections remain above censoring threshold; B, no quantified detections remain after blank censoring; mg/L, milligram per liter; n, number of sample pairs; na, not applicable; nc, not censored because constituent was detected in all samples; nd, no quantified detections remained after blank censoring; NH_4, ammonium cation; ns, not significant at 0.05 level in 2-sided test; PPW, paired Prentice-Wilcoxon; µg/L, microgram per liter; −, PPW test was not run; <, less than; ≤, less than or equal to]

| Constituent | Symbol or abbreviation | Units | Optimal censoring threshold[1] | Paired Prentice-Wilcoxon test | | | Reason no PPW test was run |
				[2]n	p-value	Sampling period with significantly higher concentration	
Aluminum	Al	µg/L	400	41.5	0.6963	ns	na
Ammonia as N	N (ammonia)	mg/L as N	0.04	44.5	<0.0001	Pre-landfall	na
Ammonia as NH_4	N (ammonium)	mg/L as NH_4	0.0515	43	<0.0001	Pre-landfall	na
Ammonia plus organic N	N (Kjeldahl)	mg/L	2.4	−	−	−	A
Arsenic	As	µg/L	40	−	−	−	A
Barium	Ba	µg/L	nc	40	0.0001	Post-landfall	na
Beryllium	Be	µg/L	10	−	−	−	A
Cadmium	Cd	µg/L	10	−	−	−	A
Calcium	Ca	mg/L	nc	42	0.0122	Post-landfall	na
Chromium	Cr	µg/L	20	40	0.3173	ns	na
Cobalt	Co	µg/L	30	−	−	−	A
Copper	Cu	µg/L	38	−	−	−	A
Iron	Fe	µg/L	500	40	0.0692	ns	na
Lead	Pb	µg/L	20	40	0.5834	ns	na
Magnesium	Mg	mg/L	nc	42	0.0024	Post-landfall	na
Manganese	Mn	µg/L	10	40	0.073	ns	na
Molybdenum	Mo	µg/L	20	40	0.0317	Post-landfall	na
Nickel	Ni	µg/L	75	−	−	−	A
Nitrogen, organic	N (organic)	mg/L	2.4	41.5	0.3173	ns	na
Nitrogen, dissolved	N (total)	mg/L	nc	41	0.8752	ns	na
Phosphorus	P	mg/L as P	0.18	42	0.0046	Pre-landfall	na
Potassium	K	mg/L	nc	42	<0.0001	Post-landfall	na
Selenium	Se	µg/L	40	−	−	−	A
Silver	Ag	µg/L	nd	−	−	−	B
Sodium	Na	mg/L	nc	42	0.0007	Post-landfall	na
Zinc	Zn	µg/L	80	40	0.9859	ns	na

[1]Lowest detection threshold that maximizes the number of quantifiable detections, minimizes the number of indeterminate samples, and has ≤7-percent indeterminate samples.

[2]Non-integer indicates data missing for member of one sample pair.

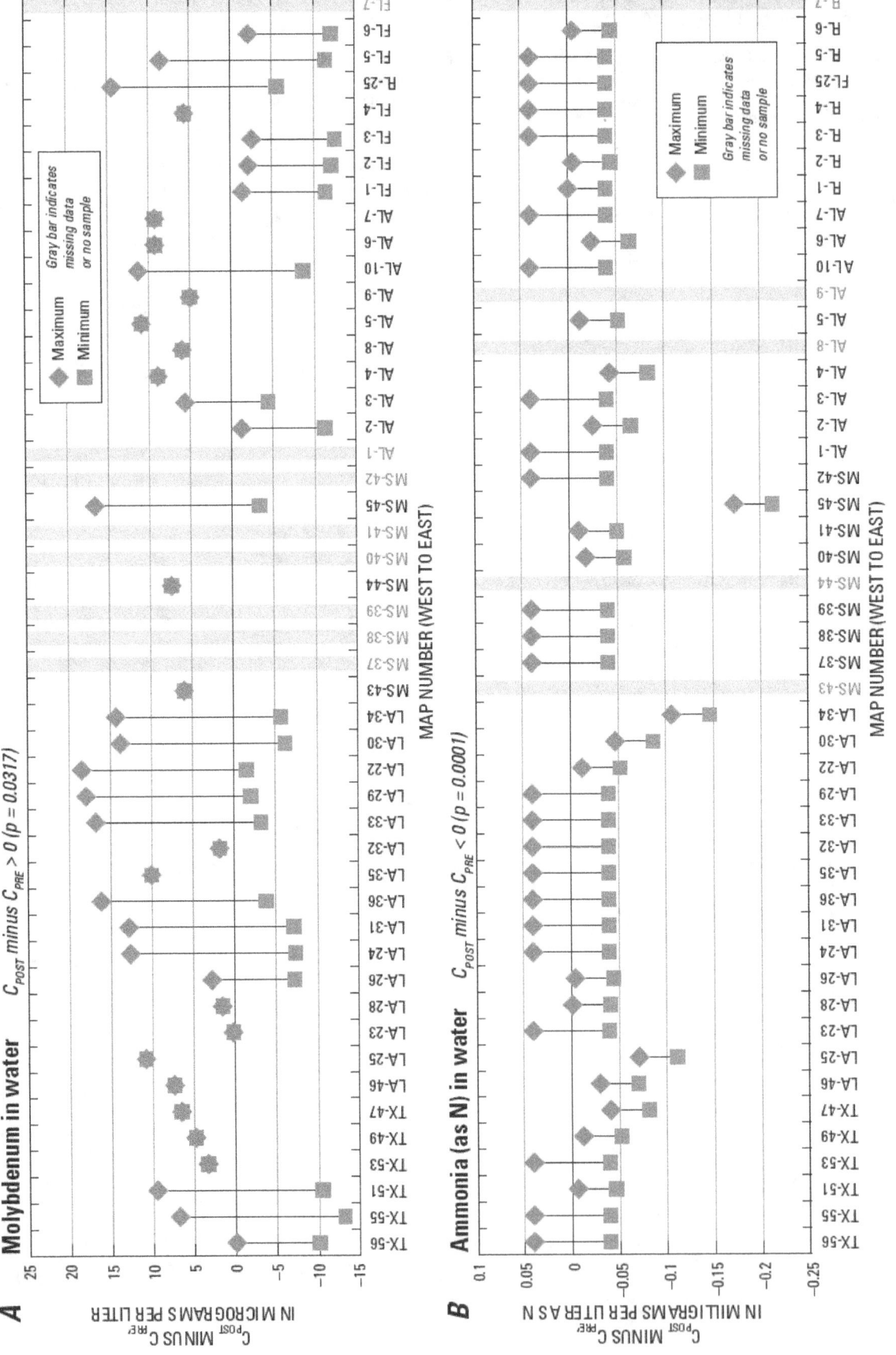

Figure 6.   For selected analytes, difference between trace and major element and nutrient concentrations in samples collected during post-landfall and pre-landfall periods (C_{post} minus C_{pre}) by the USGS at Deepwater Horizon oil spill sampling sites along the Gulf of Mexico coast, 2010, from west to east: (A) molybdenum in water, (B) ammonia as N in water, (C) lead in whole sediment, and (D) lead in the less than 63-micrometer (<63-μm) sediment fraction.

Note: < is less than; > is greater than

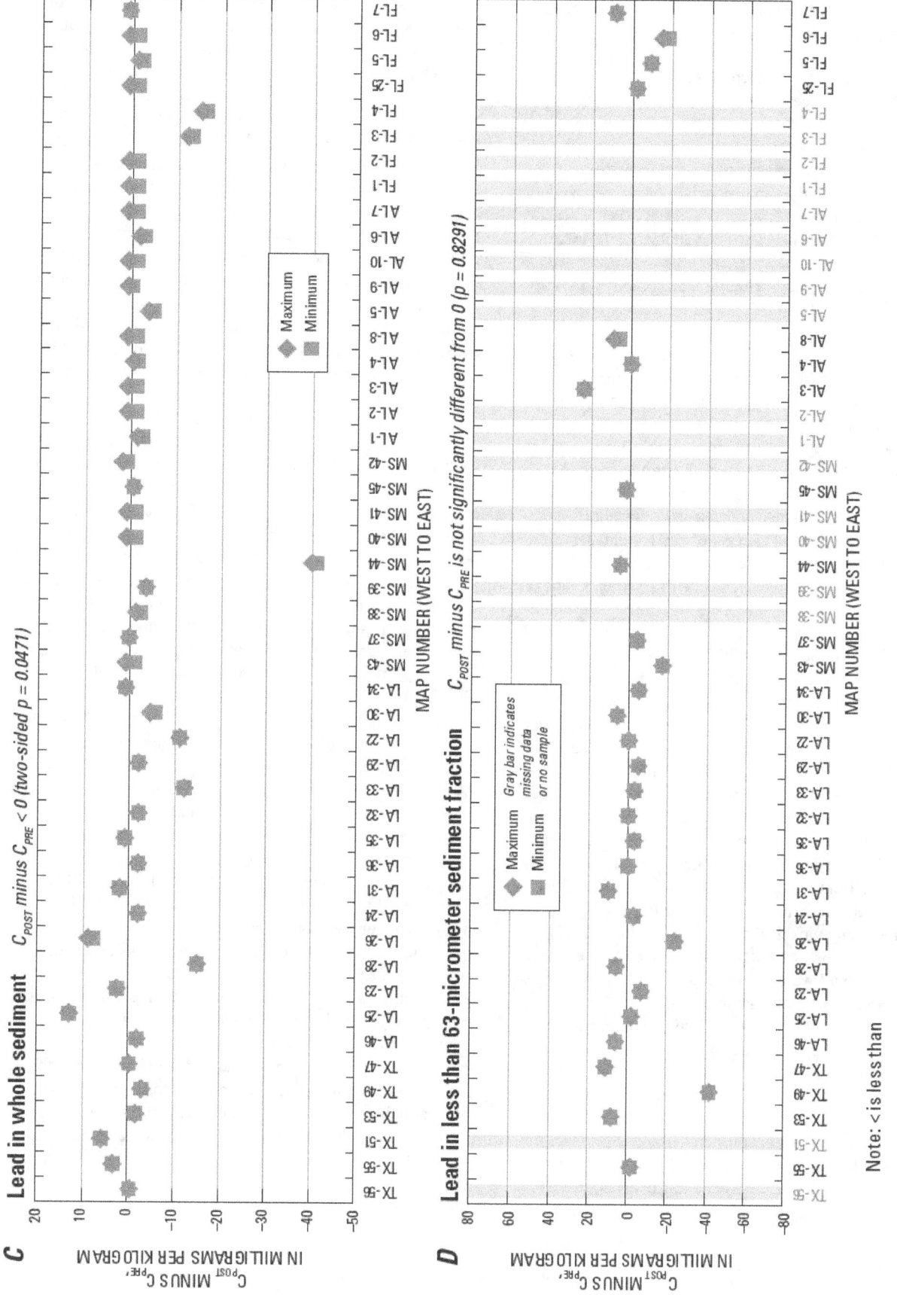

C Lead in whole sediment C_{POST} minus $C_{PRE} < 0$ (two-sided p = 0.0471)

D Lead in less than 63-micrometer sediment fraction C_{POST} minus C_{PRE} is not significantly different from 0 (p = 0.8291)

Note: < is less than

Figure 6.—Continued

Comparison with Benchmarks for Human Health and Aquatic Life

U.S. Environmental Protection Agency (2010) recommended that concentrations of nickel and vanadium in water be compared to human-health benchmarks for recreational exposure. Neither was exceeded in any water samples collected in the present study, and recreational exposure-based human-health benchmarks were not available for other trace elements.

Aquatic-life benchmarks were available for 18 trace elements in water (table 6A). Benchmarks were identified from a number of sources, including USEPA and NOAA, and included both acute and chronic marine benchmarks. As noted previously, trace-element concentrations were converted from total to dissolved concentrations by the use of marine conversion factors from the U.S. Environmental Protection Agency (2011d). Acute aquatic-life benchmarks for one or more trace elements were exceeded in 23 of 158 water samples (table 24A, appendix table 3-3), of which 22 samples were from the post-landfall period and 1 was from the pre-landfall period; these samples with observed acute benchmark exceedances represent 39 percent of post-landfall samples and 1 percent of pre-landfall samples. The elements responsible for acute benchmark exceedances were copper in all 23 samples, and zinc in 2 samples. The one pre-landfall sample with exceedances was from Louisiana, whereas post-landfall samples with exceedances were found in all five states sampled. In addition, chronic aquatic-life benchmarks were exceeded by concentrations of one or more trace elements in 74 of 158 samples, including 22 of 102 pre-landfall samples and 52 of 56 post-landfall samples; the samples with observed chronic benchmark exceedances represent 22 percent of total pre-landfall samples and 93 percent of total post-landfall samples. Overall, boron exceeded the chronic benchmarks in 50 water samples, manganese in 30, copper in 24, cobalt in 19, nickel in 7, lead in 6, barium in 3, zinc in 2, and vanadium in 1 water sample. The post-landfall sample from site LA-25, Rockefeller Refuge Beach, Louisiana, exceeded chronic benchmarks for eight trace elements, including nickel and vanadium; excluding this sample, the other post-landfall chronic benchmarks exceeded were for boron in 47 water samples, copper in 21, manganese in 11, and barium in 1 water sample.

For trace elements in water, statistical comparison of the proportion of samples exceeding aquatic-life benchmarks between the pre-landfall and post-landfall sampling periods was precluded because of the highly variable reporting levels and the large number of censored values that had reporting levels greater than the applicable benchmarks. For example, of 40 post-landfall samples in the paired-sample dataset, 1 sample exceeded the 8.1 µg/L aquatic-life benchmark for lead, 3 samples were less than this benchmark value,

16 samples were reported as censored values of less than 10 µg/L, and 20 samples were reported as censored values of less than 20 µg/L. The lead concentration could exceed the benchmark in none, some, or all of the 36 samples reported as having less than 10 or less than 20 µg/L of lead. The single benchmark exceedance observed for lead in post-landfall samples in the paired-sample dataset, therefore, represents the minimum number of exceedances of this benchmark for the 40 post-landfall samples, and the actual number of post-landfall samples with lead concentrations higher than 8.1 µg/L in this dataset could be substantially greater—theoretically, as few as 1 sample and as many as 37 of the 40 post-landfall water samples could exceed the benchmark for lead. In this sense, the uncensored benchmark-exceedance frequencies presented in this report are essentially minimum exceedance frequencies; if the analytical methods used for post-landfall samples were more sensitive, it is possible that a greater number of benchmark exceedances would have been identified. For antimony, boron, and vanadium, comparisons to benchmarks were limited because these elements were analyzed largely during only one sampling period. For arsenic, cadmium, cobalt, copper, lead, nickel, and silver, benchmark exceedance could not be ascertained for between 35 and 100 percent of samples during one or both sampling periods because concentrations were censored values at reporting levels that were higher than the applicable benchmarks. For the following analytes and sampling periods, therefore, the exceedance frequencies presented in this report could be substantially underestimated: antimony, boron, and vanadium in the pre-landfall period; arsenic, cadmium, cobalt, lead, nickel, and silver in the post-landfall period; and copper in both sampling periods. This is illustrated in figure 7, which shows the number of aquatic-life benchmark exceedances, by element and sampling period, for the 40 sample pairs in the paired-sample data set. The blue and red bars in figure 7A represent the number of observed benchmark exceedances for a given element in pre-landfall and post-landfall periods, respectively. The height of each bar in figure 7A represents the minimum number of exceedances for that element and sampling period because some samples had missing data or were censored at reporting levels too high to ascertain whether or not the benchmark was exceeded. In figure 7B, samples that are missing data (antimony, boron, and vanadium), or are censored values with reporting levels higher than the applicable benchmark, are assumed to be possible benchmark exceedances and are shown as a lighter colored segment in the stacked bar (lighter blue for pre-landfall samples and lighter red for post-landfall samples); the total height of the stacked bar represents the maximum number of exceedances possible for that element and sampling period. It is clear that several trace elements have reporting levels above the applicable benchmarks in a large number of samples, or were not analyzed in a large number of samples.

Table 24A. Summary of benchmark and baseline exceedances for trace and major elements from the Deepwater Horizon oil spill, Gulf of Mexico, 2010: benchmark exceedances in water.

[Abbreviations: B, boron; Ba, barium; Co, cobalt; Cu, copper; Mn, manganese; Ni, nickel; Pb, lead; V, vanadium; Zn, zinc; <, less than; –, no benchmarks exceeded]

State	Number of sites	Total number of samples[1]	Number of samples with benchmark exceedances			Elements that exceeded benchmarks (number of samples)	
			Human health	Aquatic life: acute	Aquatic life: chronic	Aquatic life: acute	Aquatic life: chronic
Pre-landfall sampling period							
Texas	10	12	0	0	9	–	Co (9), Pb (2), Mn (9), Ni (2)
Louisiana	15	19	0	1	9	Cu (1), Zn (1)	Ba (1), Co (9), Cu (2), Pb (3), Mn (7), Ni (4), Zn (1)
Mississippi	3	3	0	0	1	–	Co (1), Mn (1)
Alabama	9	13	0	0	3	–	B (2), Mn (1)
Florida	26	55	0	0	0	–	–
Total pre-landfall	**63**	**102**	**0**	**1**	**22**		
Post-landfall sampling period							
Texas	6	8	0	4	8	Cu (4)	B (8), Cu (4), Mn (2)
Louisiana	16	18	0	7	14	Cu (7), Zn (1)	Ba (2), B (10), Cu (7), Pb (1), Mn (7), Ni (1), V (1), Zn (1)
Mississippi	9	10	0	6	10	Cu (6)	B (10), Cu (6), Mn (2)
Alabama	10	11	0	3	11	Cu (3)	B (11), Cu (3), Mn (1)
Florida	8	9	0	2	9	Cu (2)	B (9), Cu (2)
Total post-landfall	**49**	**56**	**0**	**22**	**52**		
Total (all samples)	**112**	**158**	**0**	**23**	**74**		

[1]The total number of samples includes field replicate samples, which typically were collected from a given site on the same date.

Table 24B. Summary of benchmark and baseline exceedances for trace and major elements from the Deepwater Horizon oil spill, Gulf of Mexico, 2010: benchmark exceedances in whole sediment and national baseline exceedances in the less than 63-micrometer (µm) sediment fraction.

[Abbreviations: Al, aluminum; As, Arsenic; Ba, barium; Co, cobalt; Cr, chromium; Mn, manganese; SV, screening value; V, vanadium; –, no elements exceeded both upper SV benchmarks and national baselines; <, less than]

State	Number of sites	Total number of samples[1]	Number of whole-sediment samples in each range			Upper SV benchmarks and national baselines exceeded for the same element[5]	
			Minimal effect range[2]	Possible effect range[3]	Probable effect range[4]	Number of samples exceeding both for any element	Elements exceeding both (number of samples per element)
Pre-landfall sampling period							
Texas	10	12	0	0	12	3	As (1), Ba (2), Co (1), Mn (1)
Louisiana	15	18	0	0	18	6	As (1), Ba (3), Mn (1), V (2)
Mississippi	9	11	8	1	2	0	–
Alabama	10	10	9	1	0	0	–
Florida	26	32	25	6	1	0	–
Total pre-landfall	**70**	**83**	**42**	**8**	**33**	**9**	
Post-landfall sampling period							
Texas	6	9	0	0	9	0	
Louisiana	16	24	0	0	24	10	Al (3), Ba (9), Cr (2), Mn (1), V (3)
Mississippi	9	9	8	0	1	0	–
Alabama	10	10	10	0	0	0	–
Florida	8	8	8	0	0	0	–
Total post-landfall	**49**	**60**	**26**	**0**	**34**	**10**	
Total (all samples)	**119**	**143**	**68**	**8**	**67**	**19**	

[1]The total number of samples includes field replicate samples.

[2]All constituent concentrations were below all applicable benchmarks; there is no indication of adverse effects on benthic organisms within this range.

[3]One or more constituents exceeded a lower SV benchmark, but no elements exceeded an upper SV benchmark; adverse biological effects occasionally occur within this range.

[4]One or more constituents exceeded an upper SV benchmark; adverse biological effects frequently occur within this range.

[5]The listed elements exceeded one or more of their respective upper SV benchmarks in whole sediment, and also were enriched relative to national baseline concentrations from Horowitz and Stephens (2008).

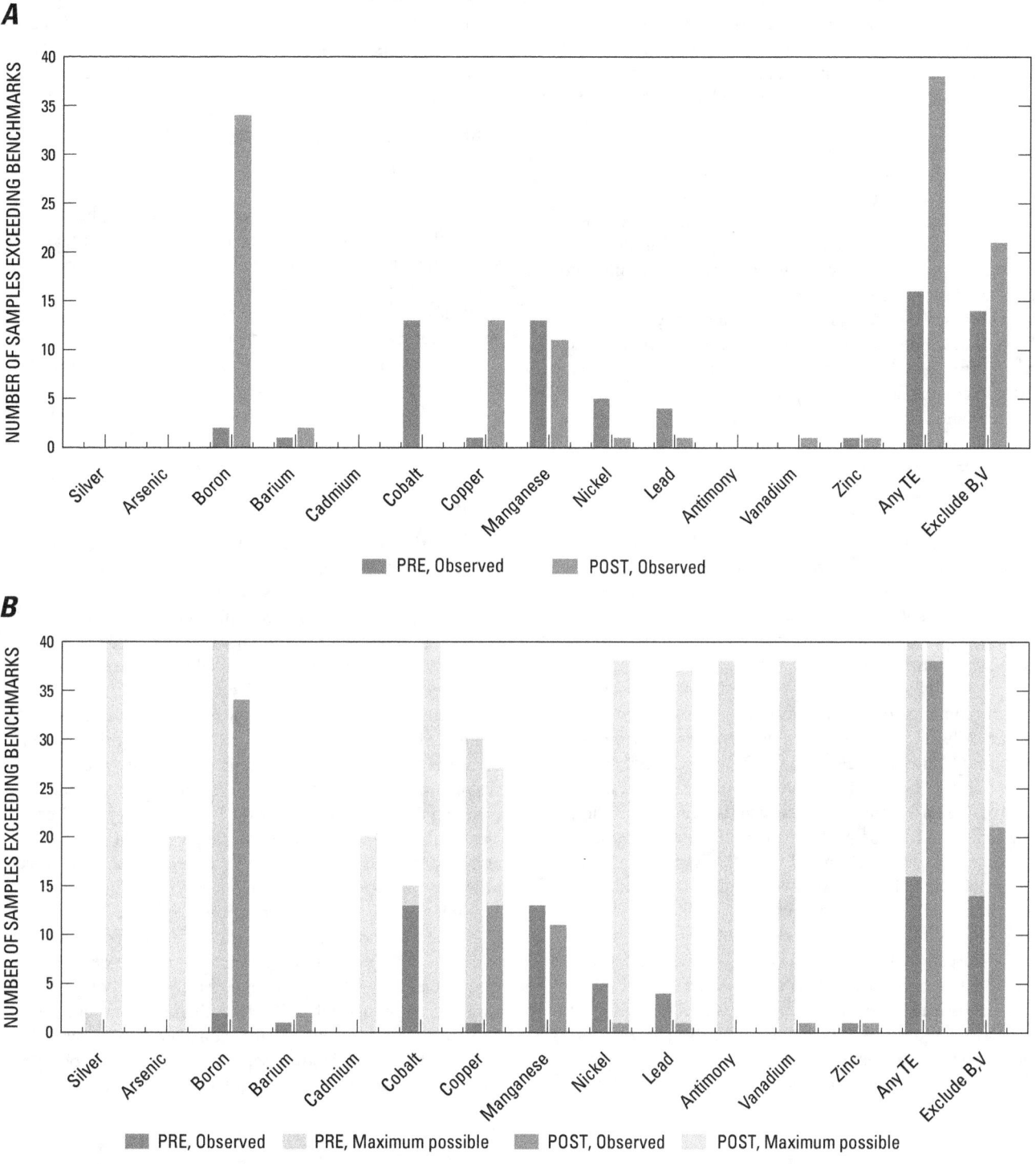

Figure 7. Number of benchmark exceedances for trace elements in water in paired pre-landfall and post-landfall samples (N = 40) from the Deepwater Horizon oil spill, Gulf of Mexico, 2010: (*A*) observed number of exceedances, which ignores censored data and represents the minimum number of exceedances; (*B*) maximum number of benchmark exceedances possible, which assumes that all samples with censored reporting levels greater than the applicable benchmark are possible exceedances. B, boron; N, number of sample pairs in the dataset; POST, post-landfall samples; PRE, pre-landfall samples; TE, trace element; V, vanadium.

Therefore, the number of benchmarks exceeded could be substantially higher than what was observed, so the benchmark exceedances observed in 47 percent of water samples represent the minimum number of exceedances for the samples collected in the present study. Furthermore, there is too much uncertainty to do statistical comparisons between sampling periods of benchmark exceedances for antimony, boron, vanadium, cobalt, copper, lead, nickel, arsenic, cadmium, and silver. For barium, manganese, and zinc, which were the only elements with sufficient data to make statistical comparisons, there were no significant differences in benchmark exceedances between paired pre-landfall and post-landfall samples (sign test, p>0.05).

Two of the elements with benchmark exceedances, boron and copper, were detected in one of four field blanks for the study, so their concentrations were blank-censored prior to comparison to benchmarks. The maximum boron concentration of 10 µg/L detected in blanks was less than 1 percent of the benchmark value of 1,200 µg/L, indicating there is reasonable certainty that measured concentrations above the benchmark were not affected by incidental contamination. For copper, however, the maximum concentration of 2.2 µg/L detected in blanks was close to the chronic and acute aquatic-life benchmarks of 3.1 and 4.8 µg/L, respectively. Therefore, measured copper concentrations were considered to be benchmark exceedances only when they exceeded 11 µg/L, or five times the maximum blank concentration.

A total of seven samples exceeded chronic aquatic-life benchmarks for nickel, vanadium, or both, which were specifically identified by U.S. Environmental Protection Agency (2011a) as relevant to the Deepwater Horizon oil spill in the GOM. Of these seven samples, six were from the pre-landfall period and exceeded the chronic benchmark for nickel, and one sample from the post-landfall period exceeded chronic benchmarks for both nickel and vanadium (appendix table 3-3). Nickel benchmark exceedance could be substantially underestimated during the post-landfall period because the reporting level of 15 to 75 µg/L was too high to ascertain whether the chronic aquatic-life benchmark of 8.2 µg/L was exceeded. Vanadium was analyzed in all post-landfall samples but in only two pre-landfall samples.

The frequency at which aquatic-life benchmarks for one or more trace elements were exceeded—47 percent—in GOM water samples indicates there is potential for toxicity to aquatic life. Because of high and variable analytical reporting levels for trace elements in water, it was not possible to do a rigorous statistical comparison of benchmark exceedances between the pre-landfall and post-landfall sampling periods.

Trace and Major Elements and Nutrients in Sediment

Trace and major elements and the nutrients phosphorus and total nitrogen were analyzed in both whole sediment and the less than 63-µm sediment fraction, by the USGS SCL for both pre-landfall and post-landfall samples. Concentrations in whole sediment were used to assess contaminant occurrence and for comparison to sediment-quality benchmarks. Concentrations in the less than 63-µm fraction of sediment were compared to national baseline concentrations in bed sediments of rivers and streams from Horowitz and Stephens (2008).

Constituent Occurrence

The detection frequencies and percentile concentrations for trace and major elements and nutrients in sediment are shown in table 25 for whole sediment and table 26 for the less than 63-µm sediment fraction. Because sediment samples were subjected to strong-acid digestion, which destroys the sediment matrix, the analyses yielded total trace-element concentrations (Horowitz and Stephens, 2008).

Detection frequencies for these constituents in whole sediment ranged from zero for thallium and uranium to over 90 percent for barium, manganese, phosphorus, sodium, strontium, and sulfur in one or both sampling periods at a common detection threshold of 0.1 mg/kg (table 25). For almost all constituents, detection frequencies in pre-landfall and post-landfall samples were separated by about 10 percent or less at their optimal censoring thresholds (table 25). As examples, the concentration distributions in pre-landfall and post-landfall whole-sediment samples are shown for calcium and lead in figures 4I and 4J. Although calcium detection frequencies above the optimum censoring threshold were similar for pre-landfall and post-landfall samples, at 67 to 68 percent, respectively, calcium concentrations appeared to be higher in post-landfall samples (fig. 4I). The opposite was true for lead, which had similar detection frequencies in

Table 25. Summary statistics for trace and major elements and nutrients in whole sediment from the Deepwater Horizon oil spill, Gulf of Mexico, 2010.

Table 26. Summary statistics for trace and major elements and nutrients in the less than 63-micrometer sediment fraction from the Deepwater Horizon oil spill, Gulf of Mexico, 2010.

These tables are presented as Microsoft© Excel spreadsheets. They can be accessed and downloaded at URL http://pubs.usgs.gov/sir/2012/5228.

both sampling periods but higher concentrations during the pre-landfall period (fig. 4J). Because whole-sediment samples were collected from the swash zone and analyzed without pre-treatment, dried sea salt could have contributed to the sodium, potassium, calcium, and magnesium concentrations measured in whole-sediment samples.

Of nutrients, phosphorus was detected above 0.1 mg/kg in all pre-landfall and post-landfall whole-sediment samples (table 25), and the highest concentrations were in two pre-landfall samples (fig. 4K). Total nitrogen was detected above its optimum threshold of 0.1 percent in 10 to 11 percent of both pre-landfall and post-landfall whole-sediment samples.

Comparison of Pre-Landfall to Post-Landfall Samples

Statistical comparison of trace- and major-element and nutrient concentrations in whole sediment was performed for 33 constituents, including total carbon and organic carbon. Six constituents showed a significant difference between pre-landfall and post-landfall samples in PPW tests (p <0.05) after censoring each element to its specific optimal censoring threshold (table 27). Concentrations were higher in post-landfall samples for calcium (fig. 4I), total carbon, sodium, and strontium, and in pre-landfall samples for lead (fig. 4J) and mercury. Using lead in whole sediment as an example, figure 6C shows the difference in concentrations when pre-landfall concentrations were subtracted from post-landfall ones ($C_{post} - C_{pre}$) at each sampling site along the GOM coast from west to east. Statistically higher lead concentrations in pre-landfall whole-sediment samples appeared to be influenced in part by a negative difference value, showing higher pre-landfall concentrations, at three sites in Louisiana, one extreme site in Mississippi, and two sites in Florida (fig. 6C). When pre-landfall concentrations were subtracted from post-landfall ones, 50 percent of sites had difference values that were negative; 23 percent had positive difference values; and 27 percent had difference values that equaled or, if the difference value was a range, included zero, so they could not be classified as definitively positive or negative.

Because trace elements tend to be concentrated in the less than 63-μm fraction of sediment, it is possible that substantially different amounts of fine material in pre-landfall and post-landfall samples could have contributed to the few significant differences in trace- and major-element concentrations that were observed. Therefore, the PPW tests were also performed on trace- and major-element and nutrient concentrations measured in the less than 63-μm fraction of sediment (table 27). Of the six elements that showed significant differences between sampling periods in whole sediment, none were significant in the less than 63-μm sediment fraction. Several factors could contribute to the lack of significant relationships in the less than 63-μm sediment data. First, the sample size was much smaller for this fraction,

which had 15 to 16 sample pairs for nutrients and 21 to 35 for other elements, compared to 35 sample pairs for nutrients and 44 to 48 for other elements in whole sediment; this reduced the power of the test for the less than 63-μm sediment fraction. Also, any dried sea salt present on whole-sediment samples would not remain in the less than 63-μm sediment fraction because the sediment samples were rinsed with deionized water during the sieving process. This would affect salts of major ions, such as calcium, sodium, and strontium. Finally, it is possible that significant differences in the constituent concentrations between post-landfall and pre-landfall samples actually were influenced by differences in the amount of fine material in these samples.

Again, by using lead as an example, the difference in post-landfall and pre-landfall concentrations ($C_{post} - C_{pre}$) at individual sampling sites for the less than 63-μm sediment fraction (fig. 6D) can be compared to the difference for whole sediment (fig. 6C). First, the most extreme difference value for lead in whole sediment (MS-44; fig. 6C) was no longer extreme in the less than 63-μm sediment fraction (fig. 6D). Both datasets had 50 percent of sites with a negative difference value when pre-landfall concentrations were subtracted from post-landfall ones; however, the less than 63-μm sediment fraction had a positive difference value at 37 percent of sites compared to 23 percent of sites for whole sediment, and difference values were indistinguishable from zero for 13 and 27 percent of sites for the less than 63-μm fraction and whole sediment, respectively. The smaller sample size for the less than 63-μm fraction is illustrated in figures 6C and 6D also. Fine sediment data were missing for several sites, especially in Florida and Alabama (fig. 6D), because the sediment samples collected at these sites had insufficient mass in the less than 63-μm fraction to run the trace-element analysis.

Overall, there was no significant difference in the percent of fine material contained in pre-landfall and post-landfall samples (table 27). To test whether site-specific differences in the percent of fine material contributed to differences in contaminant concentrations, the differences in contaminant concentrations in whole sediment between sampling periods ($C_{post} - C_{pre}$) were regressed against the differences in the percentage of sediment that was less than 63 μm (LT63) between sampling periods ($LT63_{post} - LT63_{pre}$), in both cases subtracting pre-landfall values from post-landfall values. For lead and mercury, which had significantly higher concentrations in whole sediment pre-landfall samples than in post-landfall samples, the difference in constituent concentrations was significantly (p <0.05) related to the difference in fine material between the samples. Although not conclusive, this supports the hypothesis that site-specific differences in the amount of fine material in sediment samples could have contributed to the significantly higher concentrations of lead and mercury in pre-landfall compared to post-landfall whole-sediment samples.

Table 27. Statistical comparisons of contaminant concentrations in pre-landfall samples to those in post-landfall samples from the Deepwater Horizon oil spill, Gulf of Mexico, 2010: trace and major elements and nutrients in whole sediment and in the less than 63-micrometer sediment fraction.

[Significant p-values are shaded bright yellow (p<0.01) or orange (p<0.05). Abbreviations: A, not detected in any samples; B, not detected at the optimal censoring threshold in any samples; md, data were missing for a large number of samples; n, number of samples; nc, detected in every sample, so no censoring was necessary; nd, not detected in any samples; na, not applicable; nc, detected in every sample; na, not applicable; mg/kg, milligram per kilogram; PLT63, percent of sediment sample that passes through a 63-μm sieve; post, post-landfall period; PPW, paired Prentice-Wilcoxon; pre, pre-landfall period; <, less than; –, PPW test was not run]

Constituent or parameter	Symbol or abbreviation	Units	Whole sediment					Fine (<63-micrometer) sediment				
			Optimal censoring threshold[1]	[2]n	Paired Prentice-Wilcoxon test			Optimal censoring threshold[1]	[2]n	Paired Prentice-Wilcoxon test		
					p-value	Sampling period with significantly higher concentration	Reason no PPW test was run			p-value	Sampling period with significantly higher concentration	Reason no PPW test was run
Aluminum	Al	Percent	0.1	47	0.6710	ns	na	0.3	34.5	0.5305	ns	na
Antimony	Sb	mg/kg	0.1	47	0.4194	ns	na	0.4	32.5	0.8818	ns	na
Arsenic	As	mg/kg	nc	48	0.3870	ns	na	nc	35	0.5516	ns	na
Barium	Ba	mg/kg	1	47	0.3431	ns	na	nc	35	0.3144	ns	na
Beryllium	Be	mg/kg	0.1	47	0.9432	ns	na	0.9	31	0.9940	ns	na
Cadmium	Cd	mg/kg	0.1	46.5	0.6156	ns	na	1.3	32.5	0.3173	ns	na
Calcium	Ca	Percent	0.1	47	0.0131	Post	na	0.2	34.5	0.9697	ns	na
Carbon, total	C (total)	Percent	0.1	48	0.0010	Post	na	nc	16	0.6384	ns	na
Carbon, organic	C (organic)	Percent	0.1	44	0.2478	ns	na	nc	15	0.8023	ns	na
Chromium	Cr	mg/kg	2	48	0.9243	ns	na	5	33.5	0.2190	ns	na
Cobalt	Co	mg/kg	1	47	0.7000	ns	na	10	32	0.9853	ns	na
Copper	Cu	mg/kg	1	47	0.4051	ns	na	nc	35	0.9420	ns	na
Iron	Fe	Percent	0.1	47	0.3493	ns	na	0.2	34.5	0.9574	ns	na
Lead	Pb	mg/kg	1	48	0.0471	Pre	na	3	30	0.8291	ns	na
Lithium	Li	mg/kg	1	47	0.8672	ns	na	7	34.5	0.9516	ns	na
Magnesium	Mg	Percent	0.1	47	0.8385	ns	na	nc	35	0.0349	Post	na
Manganese	Mn	mg/kg	1	48	0.4877	ns	na	nc	35	0.0274	Pre	na
Mercury	Hg	mg/kg	0.01	48	0.0450	Pre	na	0.01	21.5	0.7717	ns	na
Molybdenum	Mo	mg/kg	1	46.5	0.5787	ns	na	13	–	–	–	B
Nickel	Ni	mg/kg	1	47	0.6200	ns	na	2	34.5	0.6101	ns	na
Nitrogen, total	N (total)	Percent	0.1	35	0.5415	ns	na	nc	16	0.5893	ns	na
Phosphorus	P	mg/kg	nc	35	0.1783	ns	na	1	35	0.0624	ns	na
Potassium	K	Percent	0.1	47	0.6918	ns	na	0.6	33	0.2753	ns	na
Selenium	Se	mg/kg	0.1	46.5	0.5803	ns	na	1.2	32.5	0.2998	ns	na

Table 27. Statistical comparisons of contaminant concentrations in pre-landfall samples to those in post-landfall samples from the Deepwater Horizon oil spill, Gulf of Mexico, 2010: trace and major elements and nutrients in whole sediment and in the less than 63-micrometer (μm) sediment fraction.—Continued

[Significant p-values are shaded bright yellow (p<0.01) or orange (p<0.05). **Abbreviations**: A, not detected in any samples; B, not detected at the optimal censoring threshold in any samples; md, data were missing for a large number of samples; n, number of pre-landfall and post-landfall sample pairs; mg/kg, milligram per kilogram; na, not applicable; nc, detected in every sample, so no censoring was necessary; nd, not detected in any samples; ns, not significant at 0.05 level in two-sided test; PLT63, percent of sediment sample that passes through a 63-μm sieve; post, post-landfall period; PPW, paired Prentice-Wilcoxon; pre, pre-landfall period; <, less than; –, PPW test was not run]

Constituent or parameter	Symbol or abbreviation	Units	Whole sediment					Fine (<63-micrometer) sediment				
			Paired Prentice-Wilcoxon test					Paired Prentice-Wilcoxon test				
			Optimal censoring threshold[1]	[2]n	p-value	Sampling period with significantly higher concentration	Reason no PPW test was run	Optimal censoring threshold[1]	[2]n	p-value	Sampling period with significantly higher concentration	Reason no PPW test was run
Silver	Ag	mg/kg	0.5	46.5	0.3173	ns	na	nd	–	–	–	A
Sodium	Na	Percent	0.1	48	0.0006	Post	na	0.5	33	0.3934	ns	na
Strontium	Sr	mg/kg	1	48	0.0346	Post	na	nc	35	0.3713	ns	na
Sulfur	S	Percent	0.01	48	0.3202	ns	na	nc	35	0.4003	ns	na
Tin	Sn	mg/kg	1	47	0.2875	ns	na	13	–	–	–	B
Titanium	Ti	Percent	0.01	47.5	0.9421	ns	na	0.03	31	0.7273	ns	na
Uranium	U	mg/kg	md	–	–	–	A	600	–	–	–	B
Vanadium	V	mg/kg	1	47	0.6394	ns	na	nc	35	0.5479	ns	na
Zinc	Zn	mg/kg	1	47	0.116	ns	na	nc	35	0.3091	ns	na
Percent of sediment <63-μm	PLT63	Percent	1	46.5	0.8166	ns	na	na	na	na	na	na

Comparison with Sediment-Quality Benchmarks and National Baseline Concentrations

About 18 trace elements have one or more empirical sediment-quality benchmarks for protection of benthic organisms in whole sediment. Of 143 whole sediment samples, 67 samples (table 24B) from 28 sites exceeded one or more upper screening values for trace elements, placing these samples in the probable effect range; therefore, these samples have a high probability of adverse effects on benthic organisms. These samples included 33 of 83 pre-landfall samples and 34 of 60 post-landfall samples. Eight pre-landfall samples exceeded one or more lower screening values for trace elements, but no upper screening values, so were in the possible-effect range. The remaining 68 samples were in the minimal-effect range, indicating no adverse effects on benthic organisms would be expected. These results could be conservatively high estimates of potential toxicity because the present study measured total trace-element concentrations in sediment, rather than the bioavailable concentrations.

The trace elements with one or more upper screening-level benchmark exceedances in whole-sediment samples were barium in 66 samples, aluminum in 34, manganese in 24, vanadium in 17, cobalt in 7, arsenic in 2, and chromium in 2 samples. Trace-element concentrations exceeded one or more upper screening values in 40 percent of pre-landfall samples and 57 percent of post-landfall samples. Fisher's exact test indicated no significant difference in benchmark-exceedance frequencies between pre-landfall and post-landfall sampling periods for whole-sediment samples in the paired dataset. This was true for both upper and lower screening-value benchmarks.

There were no blank or matrix spike QC data available for trace elements in sediment. However, the QC replicate data indicate high variability in the concentrations of four elements: magnesium, mercury, sodium, and tin. Summary statistics in table 25 and benchmark comparisons in table 23 are footnoted accordingly.

Appendix table 3-4 also lists trace and major elements for which enrichment was found in the less than 63-µm sediment fraction, relative to national maximum baseline conditions. As noted previously, elements were considered to be enriched if their maximum baseline quotients exceeded 2 for samples with less than 1 percent material in the less than 63-µm sediment fraction, or 1 for all other samples. About 20 samples with less than 1 percent material in the less than 63-µm sediment fraction had insufficient material to do trace-element determinations. Of 124 samples analyzed for trace elements in the less than 63-µm sediment fraction, 81 had less than 1 percent material in the less than 63-µm fraction and were compared to the quotient threshold of 2. All but one of these samples were enriched in at least one element.

The use of national baselines to assess anthropogenic enrichment is based on the observation by Horowitz and Stephens (2008) that the upstream or underlying rock type had a minimal effect on trace- and major-element concentrations in streambed sediment nationally compared to the effects of land use or population density. There are regional differences in soil composition, however, that likely affect trace- and major-element concentrations in the less than 63-µm sediment fraction in the present study. Gustavsson and others (2001) reported total concentrations of trace and major elements in fine soil, defined as the less than 75-µm fraction of soil, across the U.S. These authors observed much lower concentrations in the less than 75-µm soil fraction for many elements in the Gulf Atlantic–Coastal Flats, which includes the Florida sites from the present study, than in many other parts of the country; this was attributed to an abundance of quartz sand in surficial soil, combined with the wet climate, which causes leaching of many elements from the upper soil horizons. These elements include arsenic, chromium, copper, lead, manganese, nickel, vanadium, and zinc. On the basis of the soil distributions observed by Gustavsson and others (2001), we would expect concentrations of these elements to be lower at Florida sites and some Alabama and Mississippi sites in the present study compared to sites in Texas and Louisiana. In fact, the concentrations of these elements reported by Gustavsson and others (2001) for most coastal soils in Florida, Alabama, and Mississippi were below the minimum baseline concentrations (that is, below the range of natural geochemical variation) in U.S. river sediment from Horowitz and Stephens (2008). In contrast, Gustavsson and others (2001) reported that soils in the Mississippi River Alluvial Plain contained the same elements at concentrations within their national baseline ranges from Horowitz and Stephens (2008); this area corresponds to the Louisiana and Texas sites in the present study. This indicates that comparison with maximum baseline concentrations will underestimate the degree of enrichment for our study sites in Florida and parts of Alabama and Mississippi, but is generally appropriate for sites in Louisiana and Texas.

The baseline exceedance results can be considered, together with upper screening-value benchmark exceedances, to identify samples that combine enrichment above baseline with potential for toxicity (table 24B and appendix table 3-4). There were 122 sediment samples with trace-element data for both whole sediment and the less than 63-µm sediment fraction. Of these, 19 samples (16 percent) exceeded upper screening-value benchmarks for, and were enriched in, one or more of these elements: barium in 14 samples, vanadium in 5, aluminum in 3, manganese in 3, arsenic in 2, chromium in 2, and cobalt in 1 sample. These samples were evenly divided between sampling periods, with 9 pre-landfall samples and 10 post-landfall samples, and were collected from 8 sites in Louisiana and 3 sites in Texas.

Contaminant Concentrations at Sites with Macondo-1 Well Oil Fingerprint Evidence

As noted previously, sediment and tarballs sampled by the USGS at 49 post-landfall sites and 69 pre-landfall sites were analyzed for diagnostic geochemical biomarkers by Rosenbauer and others (2010, 2011). In the Rosenbauer study, extracts from tarballs and from subsamples of the same composite sediment samples analyzed in the present study were compared to the chemical "fingerprint" of BP M-1 oil. The identification of M-1 well oil in the extracts was based on a combination of the interpretation of compounds identified in the mass spectra of sample extracts and a multivariate statistical analysis of the biomarker ratios by using hierarchal cluster analyses and principal component analyses.

At pre-landfall sites, residues of oil—any oil—were found in sediment from 45 of 69 sites (65 percent of sites). None of these sediment samples correlated with the M-1 oil, although a tarball collected from site FL-18 at Coco Plum Beach near Marathon, Florida, was similar to M-1 oil (Rosenbauer and others, 2011). This pre-landfall tarball sample from site FL-18 was collected on May 24, which was 6 days after NOAA reported on May 18 that a small tendril of M-1 well oil was in close proximity to the loop current (Lubchenco, 2010).

In post-landfall samples, Rosenbauer and others (2010) found at least a trace amount of oil at 44 of 49 sites (90 percent), with evidence of M-1 oil in sediment, tarballs, or both, from 19 of the 49 sampled sites (39 percent). Of 20 tarballs collected from 19 sites, all but 2 contained M-1 oil. Five of the post-landfall sites likely contained a mixture of M-1 oil plus one or more other oils. These results indicate a high incidence of oil contamination at the post-landfall sites, with direct evidence of M-1 oil in sediment, tarballs, or both, at 19 sites. These 19 sites are identified in table 1 and figure 1.

When PPW tests were run for all contaminants on the "fingerprint-sample" dataset, which was composed of pre-landfall and post-landfall samples from only those 19 sites that had M-1 oil fingerprint evidence during the post-landfall period, the results were very similar to results for the paired-sample dataset. A few analytes with significant differences when using the paired-sample dataset were no longer significant when the fingerprint-sample data subset was used. Specifically, toluene, calcium, and molybdenum in water, and calcium in sediment, were no longer significantly higher in post-landfall samples when the smaller fingerprint-sample dataset was used. Similarly, lead had significantly higher concentrations in sediment during the pre-landfall period when using the paired-sample dataset, but not when using the fingerprint-sample dataset. The loss of significance for some analytes could be a result of the much smaller sample size in the fingerprint-sample dataset, which typically had 14 to 18 sample pairs, compared to the entire dataset, which typically had 40 to 48 sample pairs. Although the 19 sites with direct evidence of M-1 oil landfall could be expected to show significantly higher contaminant concentrations

in post-landfall samples for more analytes than in the full paired-sample dataset, this was not the case; again, the small sample size of the fingerprint-sample dataset limits the power of the test. There were three analytes for which the post-landfall concentrations were significantly higher than pre-landfall concentrations in the fingerprint-sample dataset but not in the full paired-sample dataset: two alkylated PAHs (1-methylphenanthrene and C1-alkylated dibenzothiophenes) and sulfur in sediment. In terms of potential toxicity, the most important difference between the two datasets is that toluene was no longer significant in the subset of 19 sites in the fingerprint-sample dataset. The principal conclusion from the PPW analysis, however, remains unchanged—concentrations of 20 PAHs, especially alkylated PAHs, were higher overall in post-landfall samples than pre-landfall sediment samples. Of the 19 post-landfall sites with M-1 oil, 5 sites had the largest difference in post-landfall minus pre-landfall concentrations for several PAHs. These sites are Grand Isle Beach at State Park, Louisiana (LA-31); Petit Bois Island Beach, Mississippi (MS-42); and BLM-1 (AL-8), BLM-2 (AL-9), and Fort Morgan BLM-3 (AL-10) in Alabama.

Data Issues, Data Censoring, and Quality Control

Because of the nature of this project—especially the emergency timing and the involvement of multiple agencies and organizations—there were a number of data issues that had to be resolved in order to do a technically sound analysis of the resulting data. This occurred because the pre-landfall sampling had to be done soon after the oil spill, before oil made landfall, and there were not yet recommendations in place concerning what analytes should be targeted and what methods should be used. Later, between the pre-landfall and post-landfall sampling periods, changes were made to the target analyte list and the chemical analysis methods used (Operational Science Advisory Team, 2010, appendix F), and these changes improved the number of oil-related contaminants with data for the post-landfall period. Data issues faced during the data analysis required for this report included large amounts of censored data, highly variable reporting levels for a given contaminant and sampling medium, duplicate analyses of the same sample that were either verifications by the same laboratory or re-analysis by a different laboratory, systematic differences in reporting levels between pre-landfall and post-landfall sampling periods, and differences between the two sampling periods in the contaminants that were analyzed. The complexity of data types and sources also created difficulties for database management that had to be resolved before data analysis could proceed; for example, there was incomplete information on parameters, methods, and data precision from contract laboratories. The importance of database management cannot be overemphasized, and the expertise and efforts of the USGS database managers were essential to compiling a dataset of optimized and documented data quality. All of the data issues affecting this report were resolved, and are detailed in the sections on data analysis and data censoring.

One primary tool for dealing with many of these issues was strategic data censoring, which was necessary so that the data coming from different sources and representing different sampling periods, sites, or laboratories were comparable and could be evaluated on equal grounds. Unfortunately, post-laboratory data censoring resulted in loss of information for some samples.

Consistency of methods. The fact that, for some contaminants, different laboratories were used to analyze different samples contributed to a number of data issues. Different laboratories can use different methods and often have different reporting levels for the same analyte. The latter was especially problematic when there were systematic differences in reporting levels between the two sampling periods, as occurred in this study for both organic contaminants and trace elements in water. To compare results from pre-landfall and post-landfall sampling periods, data had to be censored at a common threshold, which resulted in loss of information from the sampling period with the more sensitive method. Different laboratories also had different analyte lists, and all contaminants that were determined in only one of the sampling periods had to be dropped from the comparison between sampling periods.

Sample dilution. For trace elements in water, 77 percent of samples were diluted prior to analysis because the water samples exceeded the specific conductance or total dissolved solids thresholds for analysis by ICP-OES or ICP-MS. The dilution factor used for each sample depended on the degree to which the specific conductance or total dissolved solids threshold was exceeded. Because the reporting level increased proportionally with the sample-specific dilution factor, the sample dilution process resulted in high and variable reporting levels for trace elements in water in this study. This precluded statistical comparison between pre-landfall and post-landfall samples, and in many cases censored values were higher than the applicable aquatic-life benchmarks, which made it impossible to ascertain whether benchmarks were exceeded. The use of sample dilution could have been minimized, although not eliminated altogether, by better communication with the analyzing laboratories so that operating protocols were optimized for analysis of seawater.

Quality control. Blank censoring was used to ensure that reported contaminant concentrations in environmental water samples were not affected by incidental contamination during sample collection, processing, or analysis. Because a limited number of blanks were collected during this study, a conservative approach had to be taken when censoring environmental-sample results on the basis of contamination in blanks. Detection of an analyte in any field or trip blank resulted in censoring of concentrations of that analyte in all environmental samples collected during the same sampling period. For some analytes, such as ammonia plus organic

nitrogen and benzene in water, this resulted in the censoring of all quantified detections from the post-landfall sampling period. If more blank samples had been collected, perhaps the potential for incidental contamination in an individual environmental sample could have been represented by the concentration in a single corresponding blank, and fewer samples would have been subject to blank censoring.

Target analytes. To obtain the most complete information on contaminant benchmark exceedances, water and sediment samples should be analyzed for trace elements and organic compounds, including PAHs, alkylated PAHs, and BTEX compounds. USEPA benchmarks for total PAH mixtures in water and sediment were designed to assess cumulative potential toxicity of 41 oil-related contaminants: 18 parent PAHs, 16 alkylated PAH groups, and 7 BTEX compounds (U.S. Environmental Protection Agency, 2011a and 2011b). To obtain the most complete estimate of potential PAH toxicity, all 41 target analytes that go into this benchmark should be determined. In the present study, alkylated PAHs were not analyzed in water samples or, initially, in pre-landfall sediment samples. Although alkylated PAHs can be estimated from parent PAH concentrations by using multipliers (Mount, 2010), this method can underestimate the total PAH benchmark toxic-unit value ($\sum$ESBTU or $\sum$TU) when parent PAHs are not detected. The Operational Science Advisory Team (2010; appendix table C-3) tested the efficacy of the multiplier-based estimation method by calculating toxic-unit benchmarks in two different ways for samples with a full suite of analytes measured: they compared the results obtained by using data for 16 parent PAHs plus multipliers to the results obtained by using data for all 41 analytes. Although the toxic-unit benchmarks obtained these two ways were correlated positively to each other, the relationship was not statistically significant ($p > 0.05$). In the present study, this means that $\sum$TU benchmarks for total PAH mixtures in water could be underestimated. For organic compounds in sediment, this omission was corrected by reanalyzing pre-landfall sediment samples for all 34 parent and alkylated PAHs. A second target analyte omission in the present study is that BTEX compounds included in the $\sum$ESBTU benchmark were not analyzed in sediment; therefore, $\sum$ESBTU values for sediment could be underestimated to some extent. At least for weathered oil, this low bias is likely to be minimal because the BTEX compounds are volatile and were not detected in a sample of weathered M-1 crude oil (State of Florida Oil Spill Academic Task Force, 2010) or in surface oil samples approaching the near shore environment after the spill (Atlas and Haven 2011).

These factors—use of different laboratories for pre-landfall and post-landfall sampling periods, high and variable reporting levels, missing data for analytes that should be included in benchmark calculations, and collection of only a limited number of blanks—led to difficulties in data analysis and interpretation. These are lessons learned that can be the basis for improvements in the agency response to future oil spills or similar environmental emergencies.

Summary and Conclusions

In response to the British Petroleum (BP) Deepwater Horizon Macondo-1 (M-1) oil spill on April 20, 2010, the U.S. Geological Survey (USGS) sampled beach water and sediment at 70 sites along the Gulf of Mexico (GOM) coast from May 7 to July 7, 2010, in order to establish baseline contaminant levels in potentially vulnerable locations before the oil made landfall. After the oil made landfall, a subset of 48 sites was resampled from October 4 to 14, 2010, and one new site was sampled on August 23, 2010, to assess the existence of actionable levels of M-1 oil contamination after the extensive clean-up efforts of coastal areas by BP (Wilde and Skrobialowski, 2011). This report characterizes the water and sediment chemistry in pre-landfall and post-landfall samples, evaluates whether there were significant differences between the two sampling periods, and compares measured concentrations to applicable benchmarks for human health and aquatic life.

Organics in Water

For organic contaminants in water, detection frequencies and concentrations were generally low and were similar in pre-landfall and post-landfall samples. Of the 11 compounds with enough quantified detections to statistically compare pre-landfall and post-landfall samples, concentrations were significantly higher for only one organic contaminant—toluene—primarily as a result of detections in four post-landfall samples from Florida and one from Mississippi. No samples exceeded any human-health benchmarks for organic contaminants in water, which were available for 11 compounds. Aquatic-life benchmarks, which were available for 73 compounds or mixtures of related compounds, were exceeded in only one water sample. The aquatic-life benchmarks for polycyclic aromatic hydrocarbons (PAHs) and benzene, toluene, ethylbenzene, xylene and related volatile (BTEX) compounds were exceeded in the post-landfall sample from the Mississippi River at South Pass, Louisiana (site LA-35); no exceedance was observed in the corresponding pre-landfall sample for this site.

Organics in Sediment

Most PAHs and alkylated PAHs, and a few additional semivolatile organic compounds (SVOCs), were detected in one or more samples during both pre-landfall and post-landfall periods. Nine alkylated PAHs and five parent PAHs were detected at concentrations greater than or equal to 1.5 microgram per kilogram (μg/kg) in sediment at over 20 percent of sites during one or both sampling periods, despite very low organic carbon content (for example, a median of 0.1 percent) in the sampled sediments.

Concentrations were significantly higher ($p<0.05$) in post-landfall sediment samples than pre-landfall samples for 20 of the 49 organic contaminants with enough quantified detections to make statistical comparisons, consisting of 3 PAHs and 17 alkylated PAH groups. Two analytes, naphthalene and oil and grease, had higher concentrations in pre-landfall than post-landfall samples. The same results were obtained when PAH concentrations were normalized by sediment total organic carbon (TOC), indicating that the significant differences observed were not caused simply by differences in the sediment-TOC content between the two sampling periods.

Only one sediment sample exceeded the chronic equilibrium-partitioning sediment benchmark toxic-unit concentration (ESBTU) for PAH mixtures—a pre-landfall sample from Trinity Bay near Beach City, Texas (site TX-52). This indicates that aggregate PAH concentrations were potentially toxic to benthic organisms at this site at the time of sampling. Because no post-landfall sample was collected at this site, no comparison can be made between sampling periods. Empirical benchmarks—upper screening values—for PAHs were exceeded in 27 percent of sediment samples overall, indicating a high probability of toxicity to benthic organisms at the time of sampling, although not necessarily due to PAHs. These empirical benchmarks are based on past field studies in which similar PAH concentrations in sediments were associated with toxicity (MacDonald and others, 2000); because field sediments typically contain mixtures of contaminants, however, toxicity in these studies was not necessarily due to PAHs. The percentage of sediment samples that exceeded upper screening-value benchmarks was 37 percent for post-landfall samples and 22 percent for pre-landfall samples; there was no significant difference, however, in the proportion of sediment samples that exceeded one or more benchmarks for organic contaminants in sediment between paired pre-landfall and post-landfall samples. About 70 percent of all sediment samples were below all empirical sediment-quality benchmarks for organic contaminants, indicating that no adverse effects on benthic organisms would be expected. Sediment sampled in this study typically had low organic carbon content, which could affect bioavailability and potential toxicity.

For 15 of the 17 alkylated PAHs with statistically higher concentrations in post-landfall samples, 7 sites stood out as having the largest concentration differences. For five of these seven sites, M-1 oil was identified in post-landfall sediments, tarballs, or both, on the basis of diagnostic geochemical biomarkers (Rosenbauer and others, 2010): Grand Isle Beach at State Park, Louisiana (LA-31); Petit Bois Island Beach, Mississippi (MS-42); and BLM-1, BLM-2, and Fort Morgan BLM-3 in Alabama (AL-8, AL-9, and AL-10). These results indicate that M-1 oil could have contributed to the higher PAH concentrations measured in post-landfall samples at these five sites. For the seven post-landfall sediment samples collected at these five sites, the chronic ∑ESBTU values calculated for PAH mixtures ranged from 0.2 to 0.3, and six samples, including at least one from each site, exceeded multiple upper screening-level benchmarks for total PAHs. In contrast, the nine pre-landfall sediment samples that were collected from these five sites had chronic ∑ESBTU values of less than 0.005, and no empirical screening-value benchmarks were exceeded.

Trace and Major Elements and Nutrients in Water

Detection frequencies ranged from 0 to 100 percent, depending on the element or nutrient. It was essential to censor data to a common detection threshold prior to comparing concentrations and detection frequencies for different constituents or sampling periods because reporting levels varied by constituent and by laboratory. Of the 17 trace and major elements with enough quantified values to make statistical comparisons, concentrations in water were significantly higher ($p < 0.05$) in post-landfall samples for barium, calcium, magnesium, molybdenum, potassium, and sodium. These are all elements in seawater, and barium sulfate is a standard additive in drilling mud. Ammonia and phosphorus concentrations were significantly higher ($p < 0.05$) in pre-landfall samples.

Aquatic-life benchmarks were available for 18 trace elements in water. Acute and chronic benchmarks were exceeded in 1 percent and 29 percent, respectively, of pre-landfall water samples. Post-landfall water samples exceeded acute and chronic benchmarks in 21 percent and 93 percent of samples, respectively. The elements that exceeded acute benchmarks in one or more water samples from either sampling period were copper in 23 samples and zinc in 2 samples. The elements that exceeded chronic benchmarks were boron in 50 water samples, manganese in 30, copper in 24, cobalt in 19, nickel in 7, lead in 6, barium in 3, zinc in 2, and vanadium in 1 water sample. One or more exceedances occurred in every state except Florida during the pre-landfall period, and in all five states during the post-landfall period. Of the 56 post-landfall samples, 52 exceeded one or more chronic aquatic-life benchmarks for trace elements in water, with exceedances for boron in 48 post-landfall samples, copper in 22, manganese in 12, barium in 2, and lead, nickel, vanadium, and zinc in 1 post-landfall water sample each. Because of high and variable analytical reporting levels for several trace elements in water, it was not possible to rigorously compare benchmark exceedances between the pre-landfall and post-landfall sampling periods. Overall, the frequency at which aquatic-life benchmarks for

trace elements were exceeded in GOM water samples was 47 percent, which indicates there is potential for toxicity to aquatic life. Moreover, exceedance frequencies for several trace elements could be substantially underestimated because either the element was analyzed during only one sampling period or exceedance could not be ascertained for samples that were censored at reporting levels higher than the applicable benchmark. Aquatic-life benchmark exceedance could not be ascertained for at least 35 percent of samples within a sampling period for boron and vanadium in the pre-landfall period; for arsenic, cadmium, cobalt, lead, nickel, and silver in the post-landfall period; and for copper in both sampling periods. Nickel and vanadium, which were specifically identified by U.S. Environmental Protection Agency (2011a) as relevant to the oil spill, were responsible for exceedances in only 1 of the 52 post-landfall samples with exceedances, although the results for nickel could be underestimated because of high reporting levels during the post-landfall period.

Trace and Major Elements and Nutrients in Sediment

Detection frequencies for trace and major elements and nutrients in whole sediment ranged from 0 to 100 percent, depending on the constituent, and they were similar for pre-landfall and post-landfall samples. Because sediment samples were subjected to strong acid digestion, concentrations represent total concentrations in sediment, which are defined as greater than or equal to 95 percent of the amount present.

A few trace and major elements had significant differences in concentration in whole sediment between post-landfall and pre-landfall samples; however, these differences were not significant when tests were run on the less than 63-micrometer (μm) sediment fraction. This is likely due, at least in part, to the smaller sample size of the less than 63-μm sediment-sample dataset, although other factors also could have contributed to the lack of significance in tests on the less than 63-μm fraction. Sediment samples were rinsed with water during the 63-μm sieving process, which could have removed dried sea salt present in the whole-sediment samples; if so, this would decrease concentrations of calcium, sodium, and strontium. For lead and mercury, which were significantly higher in pre-landfall than post-landfall samples for whole sediment, but not for the 63-μm sediment fraction, a larger proportion of fine material (that is, less than 63 μm) in pre-landfall sediment samples compared to post-landfall samples at some sites could have contributed to the significant difference observed for whole-sediment samples.

Empirical sediment-quality benchmarks were available for 18 trace elements in sediment. Overall, 47 percent of whole, unsieved sediment samples exceeded one or more upper screening values for trace elements (table 24B), putting these samples in the probable effect range. These samples included 33 of 83 pre-landfall samples and 34 of 60 post-landfall samples . These results could be conservatively high estimates of benchmark exceedance because they are based on measurements of total trace-element concentrations in sediment. For trace elements in whole sediment, there was no significant difference in the proportion of samples exceeding one or more aquatic-life benchmarks between the pre-landfall and post-landfall sampling periods. For the less than 63-μm sediment fraction, all but 1 of 124 samples were anthropogenically enriched in one or more trace or major elements, relative to national baseline values for U.S. streams (Horowitz and Stephens, 2008). Sixteen percent of sediment samples exceeded upper screening-value benchmarks for, and were enriched in, one or more of these elements: barium in 14 samples, vanadium in 5, aluminum in 3, manganese in 3, arsenic in 2, chromium in 2, and cobalt in 1 sample. These samples were divided evenly between the pre-landfall and post-landfall periods, and they were collected from 8 sites in Louisiana and 3 sites in Texas. However, because many trace elements have lower concentrations in soils from Florida, Alabama, and Mississippi than in soils from Louisiana and Texas (Gustavsson and others, 2001), the baseline comparison analysis probably underestimates the degree of enrichment at Florida, Alabama, and Mississippi sites.

Comparison of Pre-Landfall to Post-Landfall Samples

Considering all the information evaluated in this report, there were significant differences between post-landfall and pre-landfall samples for PAH concentrations in sediment. With a few exceptions, pre-landfall and post-landfall samples did not differ significantly in concentrations or benchmark exceedances for most organics in water or trace elements in sediment. The one exception is toluene, which had significantly higher concentrations in post-landfall than pre-landfall water samples, although this difference was not necessarily related to landfall M-1 oil. Toluene is volatile and was not detected in weathered M-1 crude oil (State of Florida Oil Spill Academic Task Force, 2010) or in surface-oil samples approaching the near shore environment after the spill (Atlas and Haven, 2011). For trace elements in water, aquatic-life benchmarks were exceeded in 47 percent of samples overall, but the high and variable analytical reporting levels

precluded statistical comparison of benchmark exceedances between sampling periods. Of the organic contaminants in sediment, 3 parent PAHs and 17 alkylated PAH groups had significantly higher concentrations in post-landfall samples than in pre-landfall samples. Concentrations above the upper screening-value benchmarks put 37 percent of post-landfall samples and 22 percent of pre-landfall samples in the probable-effect range. However, the proportion of samples exceeding empirical upper screening-value benchmarks for PAHs in sediment were not significantly different in paired post-landfall and pre-landfall samples.

For 15 of the 17 alkylated PAHs with statistically higher concentrations in post-landfall samples, the greatest concentration differences were observed at seven sites. These results corroborate the results of Rosenbauer and others (2010), who found diagnostic geochemical evidence of Deepwater Horizon M-1 oil in post-landfall sediment and tarballs from five of these seven sites. The five sites are Grand Isle Beach at State Park, Louisiana (LA-31); Petit Bois Island Beach, Mississippi (MS-42); and BLM-1, BLM-2, and Fort Morgan BLM-3 in Alabama (AL-8, AL-9, and AL-10).

Acknowledgements

The USGS initiated the pre-landfall sampling and analytical study, and additional funding was later provided by the U.S. Coast Guard for both the pre-landfall and post-landfall work. We gratefully acknowledge the efforts of the staffs of the USGS Water Science Centers in Alabama, Florida, Georgia, Louisiana, Mississippi, and Texas in sample collection and processing. We thank Marge Davenport, Charles Demas, Yvonne Stoker, Donna Myers, and Terry Schertz for their coordination efforts; Franceska Wilde and Stan Skrobialowski for development of the sampling protocol and other assistance; and Gary Cottrell for his efforts in coordinating the laboratory contracts. We thank Arthur Horowitz, Geoffrey Plumlee, and Ruth Wolf for data and technical assistance. We also thank Robert Burgess, Christopher Ingersoll, Donald MacDonald, and David Mount for technical guidance on benchmarks. We are grateful to four anonymous reviewers for their thoughtful reviews of an earlier version of the manuscript that greatly improved this report. We thank David Strong and Gregory Wetherbee for their work on the map in this report, and Wes Stone for technical assistance. We also thank the USGS Tacoma Publishing Service Center, and Carol Sanchez, Robin Miller, Larry Schneider, and Susan Davis of the USGS Sacramento Publishing Service Center, for their support in the publication of this report.

References Cited

American Petroleum Institute, 2003, High Production Volume Chemical Challenge Program, test plan, crude oil category: Submitted to the U.S. Environmental Protection Agency, November 21, 2003: American Petroleum Institute Petroleum HPV Testing Group, accessed August 12, 2011, at URL http://www.petroleumhpv.org/docs/crude_oil/2011_jan14_Crude%20oil%20category%20Final%20CAD%20-%2014%20January%202011.pdf

American Petroleum Institute, 2011, High Production Volume Chemical Challenge Program, crude oil category, category assessment document: Submitted to the U.S. Environmental Protection Agency, January 14, 2011: American Petroleum Institute Petroleum HPV Testing Group, accessed September 23, 2011, at URL http://www.petroleumhpv.org/docs/crude_oil/2011_jan14_Crude%20oil%20category%20Final%20CAD%20-%2014%20January%202011.pdf

Apodaca, L.E., Mueller, D.K., and Koterba, M.T., 2006, Review of trace element blank and replicate data collected in ground and surface water for the National Water-Quality Assessment Program, 1991–2002: U.S. Geological Survey Scientific Investigations Report 2006-5093, accessed September 23, 2011, at URL http://pubs.usgs.gov/sir/2006/5093/sir_2006-5093.pdf

Argonne National Laboratory, ChevronTexaco, and Marathon, 2012, Drilling waste management information system: Fact sheet–Using muds and additives with lower environmental impacts: Washington, D.C., U.S. Department of Energy's Natural Gas & Oil Technology Partnership, accessed June 11, 2012, at URL http://web.ead.anl.gov/dwm/techdesc/lower/index.cfm

Atlas, R.M. and Hazen, T.C., 2011, Oil biodegradation and bioremediation: A tale of the two worst spills in U.S. history: Environmental Science and Technology, v. 45, no. 16, p. 6709–6715.

Australian and New Zealand Environment and Conservation Council, 2000, Aquatic Ecosystems, chap. 3 in The Guidelines, v.1 of Australian and New Zealand Guidelines for Fresh and Marine Water Quality, accessed January 20, 2011, at URL http://www.mincos.gov.au/_data/assets/pdf_file/0019/316126/wqg-ch3.pdf

Barrick, Robert; Becker, Scott; Brown, Lorraine; Beller, Harry; and Pastorok, Robert, 1988, Sediment quality values refinement: Vol. 1-1988, Update and evaluation of Puget Sound AET: Bellevue, WA, PTI Environmental Services, prepared for U.S. Environmental Protection Agency, Region 10, Seattle, WA, 193 p.

British Columbia Ministry of Environment, 2010, Water quality guideline reports: Environmental Protection Division, British Columbia Ministry of Environment, accessed April 20, 2011, at URL http://www.env.gov.bc.ca/wat/wq/wq_guidelines.html

Buchman, M.F., 2008, NOAA Screening quick reference tables: Seattle, WA, National Oceanic and Atmospheric Administration, Office of Response and Restoration Division Report 08-1, accessed November 22, 2010, current URL: http://archive.orr.noaa.gov/book_shelf/122_NEW-SQuiRTs.pdf

Burgess, R.M., Ahrens, M.J., and Hickey, C.W., 2003, Geochemistry of PAHs in aquatic environments: source, persistence and distribution, in Douben, P.E.T., ed., PAHs: An Ecotoxicological Perspective: Sussex, U.K., John Wiley & Sons Ltd., 376 p.

Canadian Council of Ministers of the Environment, 1995, Protocol for the derivation of Canadian sediment quality guidelines for the protection of aquatic life: Ottawa, Canada, Environment Canada, Guidelines Division, Technical Secretariat of the CCME task Group on Water Quality Guidelines, Canadian Council of Ministers of the Environment, CCME EPC-98E, [Reprinted in Canadian Council of Ministers of the Environment, ed., 1999, Canadian environmental quality guidelines, chap. 6], accessed March 31, 2011, at URL http://www.ccme.ca/assets/pdf/pn_1176_e.pdf

Canadian Council of Ministers of the Environment, 1999, Introduction, in Canadian water quality guidelines for the protection of aquatic life, in Canadian Environmental Quality Guidelines, 1999: Winnipeg, Manitoba, Canada, Canadian Council of Ministers of the Environment, accessed September 23, 2011, at URL http://ceqg-rcqe.ccme.ca/download/en/312

Canadian Council of Ministers of the Environment, 2001, Introduction, in Canadian sediment quality guidelines for the protection of aquatic life, in Canadian Environmental Quality Guidelines, 1999: Winnipeg, Manitoba, Canada, Canadian Council of Ministers of the Environment, accessed April 20, 2011, at URL http://ceqg-rcqe.ccme.ca/download/en/317

Canadian Council of Ministers of the Environment, 2011, Canadian environmental quality guidelines, Summary table: Canadian Council of Ministers of the Environment (CCME), accessed April 20, 2011, at URL http://st-ts.ccme.ca/

Connor, B.F., Rose, D.L., Noriega, M.C., Murtagh, L.K., and Abney, S.R., 1998, Methods of analysis by the U.S. Geological Survey National Water Quality Laboratory–Determination of 86 volatile organic compounds in water by gas chromatography/mass spectrometry, including detections less than reporting limits: U.S. Geological Survey Open-File Report 97-829, accessed April 14, 2011, at URL http://wwwnwql.cr.usgs.gov/USGS/pubs/OFR/OFR-97-829.pdf

Continental Shelf Associates, Inc., 1997, Gulf of Mexico Produced Water Bioaccumulation Study, Executive summaries, prepared for Offshore Operators Committee, New Orleans, LA, accessed August 6, 2012, at URL http://www.deq.louisiana.gov/portal/Portals/0/permits/lpdes/pdf/EPA%20Platform_Survey_report.pdf

Di Toro, D.M., Zarba, C.S., Hansen, D.J., Berry, W.J., Swartz, R.C., Cowan, C.E., Pavlou, S.P., Allen, H.E., Thomas, N.A., and Paquin, P.R., 1991, Technical basis for the equilibrium partitioning method for establishing sediment quality criteria: Environmental Toxicology and Chemistry, v. 11, p. 1541–1583.

Field, L.J., MacDonald, D.D., Norton, S.B., Ingersoll, C.G., Severn, C.G., Smorong, Dawn, and Lindskoog, Rebekka, 2002, Predicting amphipod toxicity from sediment chemistry using logistic regression models: Environmental Toxicology and Chemistry, v. 21, no. 9, p. 1993–2005.

Fishman, M.J., ed., 1993, Methods of analysis by the U.S. Geological Survey National Water Quality Laboratory–Determination of inorganic and organic constituents in water and fluvial sediments: U.S. Geological Survey Open-File Report 93-125, 217 p., accessed September 23, 2011, at URL http://pubs.er.usgs.gov/publication/ofr93125

Fishman, M.J., and Friedman, L.C., 1989, Methods for determination of inorganic substances in water and fluvial sediments: U.S. Geological Survey Techniques of Water-Resources Investigations, book 5, chap. A1, 545 p.

Garbarino, J.R., 1999, Methods of analysis by the U.S. Geological Survey National Water Quality Laboratory–Determination of dissolved arsenic, boron, lithium, selenium, strontium, thallium, and vanadium using inductively coupled plasma-mass spectrometry: U.S. Geological Survey Open-File Report 99-093, 31 p., accessed September 18, 2011, at URL http://pubs.usgs.gov/of/1999/0093/report.pdf

Garbarino, J.R., Kanagy, L.K., and Cree, M.E., 2006, Determination of elements in natural-water, biota, sediment, and soil samples using collision/reaction cell inductively coupled plasma–mass spectrometry: U.S. Geological Survey Techniques and Methods, book 5, sec. B, chap. 1, 88 p., accessed September 23, 2011, at URL http://pubs.usgs.gov/tm/2006/tm5b1/PDF/TM5-B1.pdf

Garbarino, J.R., and Struzeski, T.M., 1998, Methods of analysis by the U.S. Geological Survey National Water Quality Laboratory–Determination of elements in whole-water digests using inductively coupled plasma–optical emission spectrometry and inductively coupled plasma–mass spectrometry: U.S. Geological Survey Open-File Report 98-165, 101 p., accessed September 23, 2011, at URL http://wwwnwql.cr.usgs.gov/USGS/pubs/OFR/OFR-98-165.pdf

Georgia Coastal Research Council, 2010, Oil spill summit II: Chemical considerations: Jacksonville, FL, University of North Florida, South Atlantic Sea Grant, 15 p., accessed August 12, 2011, at URL http://www.gcrc.uga.edu/PDFs/SA_SG_oil_summit2_report.pdf

Gries, T.H., and Waldow, K.H., 1994, Progress re-evaluating Puget Sound apparent effects thresholds (AETs): 1994 Amphipod and Echinoderm Larval AETs, vol. 1: Olympia, Washington, Washington Department of Ecology, 88 p.

Gustavsson, N., Bølvike, B., Smith, D.B., and Severson, R.C., 2001, Geochemical landscapes of the conterminous United States–New map presentations for 22 elements: U.S. Geological Survey Professional Paper 1648, 38 p., accessed August 17, 2011, at URL http://pubs.usgs.gov/pp/p1648/

Helsel, D.R., 2005, Nondetects and data analysis: statistics for censored environmental data: Hoboken, NJ, Wiley-Interscience, 268 p.

Helsel, D.R., and Hirsch, R.M., 2002, Statistical methods in water resources, chap. A3, Book 4 in Techniques of Water Resources Investigations: U.S. Geological Survey, TWRI 4-A3, 522 p., accessed November 15, 2011, at URL http://pubs.usgs.gov/twri/twri4a3/

Horowitz, A.J., and Stephens, V.C., 2008, The effects of land use on fluvial sediment chemistry for the conterminous U.S.–Results from the first cycle of the NAWQA Program: Trace and major elements, phosphorus, carbon, and sulfur: Science of the Total Environment, v. 400, p. 290–314.

Hyland, J.L., Balthis, W.L., Engle, V.D., Long, E.R., Paul, J.F., Summers, J.K., and Van Dolah, R.F., 2003, Incidence of stress in benthic communities along the U.S. Atlantic and Gulf of Mexico Coasts within different ranges of sediment contamination from chemical mixtures: Environmental Monitoring and Assessment, v. 81, p. 149–161.

Ingersoll, C.G., MacDonald, D.D., Wang, N., Crane, J.L., Field, L.J., Haverland, P.S., Kemble, N.E., Lindskoog, R.A., Severn, C.G., Smorong, D.E., 2001, Predictions of sediment toxicity using consensus-based freshwater sediment quality guidelines. Archives of Environmental Contamination and Toxicology, v. 41, p. 8–21.

Iqbal, Javed, Overton, E.B., and Gisclair, David, 2008, Polycyclic aromatic hydrocarbons in Louisiana rivers and coastal environments: source fingerprinting and forensic analysis: Environmental Forensics, v. 9, no. 1, p. 63–74.

Leung, K.M.Y.; Bjorgesaeter, Anders; Gray, J.S.; Li, W.K.; Lui, G.C.S.; Wang, Yuan; and Lam, P.K.S., 2005, Deriving sediment quality guidelines from field-based species sensitivity distributions: Environmental Science and Technology, v. 39, p. 5148–5156.

Long, E.R., MacDonald, D.D., Smith, S.L., and Calder, F.D., 1995, Incidence of adverse biological effects within ranges of chemical concentrations in marine and estuarine sediments: Environmental Management, v. 19, no. 1, p. 81–97.

Long, E.R., and Morgan, L.G., 1991, The potential for biological effects of sediment-sorbed contaminants tested in the National Status and Trends Program: Seattle, Washington, National Oceanic and Atmospheric Administration Technical Memorandum NOS OMA 52, 235 p.

Lorenz, D.L., Ahearn, E.A., Carter, J.M., Cohn, T.A., Danchuk, W.J., Frey, J.W., Helsel, D.R., Lee, K.E., Leeth, D.C., Martin, J.D., McGuire, V.L., Neitzert, K.M., Robertson, D.M., Slack, J.R., Starn, Jeffrey, Vecchia, A.V., Wilkison, D.H., and Williamson, J.E., 2011, USGS library for S-PLUS for Windows–Release 4.0: U.S. Geological Survey Open-File Report 2011-1130.

Lubchenco, Jane, 2010, Transcript loop current conference call May 18, 2010, Deepwater Horizon response, National Oceanic and Atmospheric Administration press briefing: accessed November 28, 2012., at URL http://www.restorethegulf.gov/release/2010/05/19/transcript-loop-current-conference-call-may-18-2010

MacDonald, D.D., Carr, R.S., Calder, F.D., Long, E.R., and Ingersoll, C.G., 1996, Development and evaluation of sediment quality guidelines for Florida coastal waters: Ecotoxicology, v. 5, no. 4, p. 253–278.

MacDonald, D.D., Ingersoll, C.G., and Berger, T.A., 2000, Development and evaluation of consensus-based sediment quality guidelines for freshwater ecosystems: Archives of Environmental Contamination and Toxicology, v. 39, p. 20–31.

Martin, J.D., 2002, Variability of pesticide detections and concentrations in field replicate water samples collected for the National Water-Quality Assessment Program, 1992–97: U.S. Geological Survey Water-Resources Investigations Report 2001-4178, 84 p., accessed September 23, 2011, at URL http://pubs.usgs.gov/wri/2001/wri01_4178/pdf/wri01-4178.pdf

Meays, Cindy, 2010, Derivation of water quality guidelines to protect aquatic life in British Columbia: Science and Information Branch, Ministry of Environment, 32 p., accessed April 28, 2011, at URL http://www.env.gov.bc.ca/wat/wq/pdf/wq-derivation.pdf

Mount, Dave, 2010, Explanation of PAH benchmark calculations using EPA PAH ESB approach: U.S. Environmental Protection Agency, 6 p., available online at URL http://www.epa.gov/bpspill/water/explanation-of-pah-benchmark-calculations-20100622.pdf

Mueller, D.K., and Titus, C.J., 2005, Quality of nutrient data from streams and ground water sampled during water years 1992–2001: U.S. Geological Survey Scientific Investigations Report 2005–5106, 27 p., accessed September 23, 2011, at URL http://pubs.usgs.gov/sir/2005/5106/pdf/sir2005-5106.pdf

Nadkarni, R.A., 1991, A review of modern instrumental methods of elemental analysis of petroleum related material: Part 1–occurrence and significance of trace metals in petroleum and lubricants, in Nadkarni, R.A., ed., Modern instrumental methods of elemental analysis of petroleum products and lubricants: Philadelphia, American Society of Testing and Materials, ASTM STP 1109, p. 5–18.

Nagpal, N.K., Pommen, L.W., and Swain, L.G., 2006, A compendium of working water quality guidelines for British Columbia: Science and Information Branch, Ministry of Environment, accessed April 28, 2011, at URL http://www.env.gov.bc.ca/wat/wq/BCguidelines/working.html

National Oceanic and Atmospheric Administration, 2010, Deepwater Horizon / BP oil spill response, accessed June 6, 2012, at URL http://archive.orr.noaa.gov/dwh.php?entry_id809#downloads

Neff, J.M., 2002, Bioaccumulation in marine organisms: Effect of contaminants from oil well produced water: Elsevier Inc., San Diego, Calif, 452 p.

Neff, J., Lee, K., and DeBlois, E.M., 2011, Produced water: overview of composition, fates, and effects, in Lee, K. and Neff, J., eds. Produced water: environmental risks and advances in mitigation technologies: Springer, New York, NY, p. 3–56.

Operational Science Advisory Team, 2010, Summary report for sub-sea and sub-surface oil and dispersant detection–Sampling and monitoring: New Orleans, Unified Area Command, prepared for the U.S. Coast Guard, December 17, 2010, variously paged, accessed June 2, 2011, at URL http://www.restorethegulf.gov/sites/default/files/documents/pdf/OSAT_Report_FINAL_17DEC.pdf90

Operational Science Advisory Team, 2011, Summary report for sub-sea and sub-surface oil and dispersant detection: Ecotoxicity addendum: prepared for the U.S. Coast Guard, July 8, 2011, 35 p., accessed September 9, 2011, at URL http://www.restorethegulf.gov/sites/default/files/u306/FINAL%20OSAT%20Ecotox%20Addendum.pdf

Patton, C.J., and Kryskalla. J.R., 2003, Methods of Analysis by the U.S. Geological Survey National Water Quality Laboratory–Evaluation of alkaline persulfate Digestion as an alternative to Kjeldahl digestion for determination of total and dissolved nitrogen and phosphorus in water: U.S. Geological Survey Water-Resources Investigations Report 03-4174, 33 p.

Patton, C.J., and Truitt, E.P., 1992, Methods of analysis by the U.S. Geological Survey National Water Quality Laboratory–Determination of total phosphorus by a Kjeldahl digestion method and an automated colorimetric finish that includes dialysis: U.S. Geological Survey Open-File Report 92-146, 39 p., accessed September 23, 2011, at URL http://pubs.usgs.gov/of/1992/0146/report.pdf

Patton, C.J., and Truitt, E.P., 2000, Methods of analysis by the U.S. Geological Survey National Water Quality Laboratory–Determination of ammonium plus organic nitrogen by a Kjeldahl digestion method and an automated photometric finish that includes digest cleanup by gas diffusion: U.S. Geological Survey Open-File Report 00-170, 31 p., accessed September 23, 2011, at URL http://pubs.usgs.gov/of/2000/0170/report.pdf

Reddy, C.M., Arey, S.J., Seewald, J.S., Sylva, S.P., Lemkau, K.L., Nelson, R.K., Carmichael, C.A., McIntyre, C.P., Fenwick, J., Ventura, G.T., Van Mooy, B.A.S., and Camilli, R., 2012. Composition and fate of gas and oil released to the water column during the Deepwater Horizon oil spill. Proceedings of the National Academy of Sciences, www.pnas.org/cgi/doi/10.1073/pnas.1101242108

Rosenbauer, R.J., Campbell, P.L., Lam, A., Lorenson, T.D., Hostettler, F.D., Thomas, B., and Wong, F.L., 2010, Reconnaissance of Macondo-1 well oil in sediment and tarballs from the Northern Gulf of Mexico shoreline, Texas to Florida: U.S. Geological Survey Open-File Report 2010-1290, 22 p., accessed August 4, 2011, at URL http://pubs.usgs.gov/of/2010/1290/of2010-1290.pdf

Rosenbauer, R.J., Campbell, P.L., Lam, Angela, Lorenson, T.D., Hostettler, F.D., Thomas, Burt, and Wong, F.L., 2011, Petroleum hydrocarbons in sediment from the Northern Gulf of Mexico shoreline, Texas to Florida: U.S. Geological Survey Open-File Report 2011-1014, 22 p., accessed April 6, 2011, at URL http://pubs.usgs.gov/of/2011/1014/of2011-1014.pdf

SAS Institute Inc., 2009a, Base SAS 9.2 procedures guide: Cary, NC, SAS Institute Inc, 1,680 p.

SAS Institute Inc., 2009b, SAS/Stat 9.2 user's guide: Cary, NC, SAS Institute Inc, 7,869 p.

State of Florida Oil Spill Academic Task Force, 2010, Description of the MC252 crude oil: Florida, Florida State University, accessed August 11, 2011, at URL http://oilspill.fsu.edu/images/pdfs/mc-252crude-oil-desc.pdf

TIBCO Software, Inc., 2008, TIBCO Spotfire S+® 8.1 for Windows®, [Computer Software], available online at URL http://spotfire.tibco.com/en.aspx

Turekian, K.K., 1968, Oceans: Engelwood Cliffs, NJ, Prentice-Hall, 120 p.

Unified Area Command, 2010, Deepwater Horizon MC 252 Response–Strategic plan for the sub-sea and sub-surface oil and dispersant detection, sampling and monitoring: Unified Area Command, 95 p.

U.S. Environmental Protection Agency, 1986 (revised December 1996), Test methods for evaluating of solid waste, physical/chemical methods: Office of Solid Waste and Emergency Response, U.S. Environmental Protection Agency, EPA-SW-846 (3rd ed.), subsequent updates, variously paged, accessed August 1, 2012, at URLs http://www.caslab.com/EPA-Methods/PDF/8270c.pdf, http://www.epa.gov/wastes/hazard/testmethods/sw846/pdfs/6010c.pdf, http://www.epa.gov/wastes/hazard/testmethods/sw846/pdfs/7470a.pdf, http://www.caslab.com/EPA-Methods/PDF/8015b.pdf, http://www.epa.gov/wastes/hazard/testmethods/sw846/pdfs/8015c.pdf, http://www.epa.gov/osw/hazard/testmethods/sw846/pdfs/8260b.pdf, http://www.caslab.com/EPA-Methods/PDF/8270c.pdf, http://www.epa.gov/wastes/hazard/testmethods/sw846/pdfs/8270d.pdf

U.S. Environmental Protection Agency, 1989, Data evaluation, chap. 5 in Risk assessment guidance for Superfund: Volume I–Human health evaluation manual (Part A): Washington, D.C., Office of Emergency and Remedial Response, U.S. Environmental Protection Agency Report EPA/540/1-89/002, 30 p., accessed September 23, 2011, at URL http://www.epa.gov/oswer/riskassessment/ragsa/pdf/ch5.pdf

U.S. Environmental Protection Agency, 1996, Ecotox thresholds, EPA EcoUpdate, v. 3, no. 2, EPA 540/F-95/038, accessed January 27, 2011, at URL http://www.epa.gov/oswer/riskassessment/ecoup/pdf/v3no2.pdf

U.S. Environmental Protection Agency, 1998, Method 9071B–n-Hexane extractable material (HEM) for sludge, sediment, and solid samples: Method 9071B, accessed on September 15, 2011, at URL http://www.epa.gov/epawaste/hazard/testmethods/sw846/pdfs/9071b.pdf

U.S. Environmental Protection Agency, 1999, Method 1664, revision A: *N*-hexane extractable material (HEM; oil and grease) and silica gel treated *N*-hexane extractable material (SGT-HEM; non-polar material) by extraction and gravimetry, February 1999: Office of Water, U.S. Environmental Protection Agency, EPA 821/R-98-002, 28 p.

U.S. Environmental Protection Agency, 2002, Procedures for the derivation of equilibrium partitioning sediment benchmarks (ESBs) for the protection of benthic organisms: PAH mixtures: Office of Research and Development, U.S. Environmental Protection Agency, EPA-600-R-02-013.

U.S. Environmental Protection Agency, 2004, The incidence and severity of sediment contamination in surface waters of the United States, National Sediment Quality Survey (2d ed.): Office of Science and Technology, U.S. Environmental Protection Agency, EPA 823-R-04-007, accessed September 23, 2011, at URL http://water.epa.gov/polwaste/sediments/cs/report2004_index.cfm

U.S. Environmental Protection Agency, 2008, Procedures for the derivation of equilibrium partitioning Sediment Benchmarks (ESBs) for the protection of benthic organisms: Compendium of Tier 2 values for nonionic organics: U.S. Environmental Protection Agency, EPA/600/R-02/016, accessed September 23, 2011, at URL http://www.epa.gov/nheerl/download_files/publications/ESB_Compendium_v14_final.pdf

U.S. Environmental Protection Agency, 2009, National Recommended Water Quality Criteria: U.S. Environmental Protection Agency, accessed September 23, 2011, at URL http://www.epa.gov/ost/criteria/wqctable/

U.S. Environmental Protection Agency, 2010, Human health benchmarks for chemicals in water, EPA response to BP spill in the Gulf of Mexico: U.S. Environmental Protection Agency, accessed January 27, 2011, at URL http://www.epa.gov/bpspill/health-benchmarks html

U.S. Environmental Protection Agency, 2011a, Water quality benchmarks for aquatic life, EPA response to BP spill in the Gulf of Mexico: U.S. Environmental Protection Agency, accessed January 27, 2011, at URL http://www.epa.gov/bpspill/water-benchmarks html

U.S. Environmental Protection Agency, 2011b, Sediment benchmarks for aquatic life, EPA response to BP spill in the Gulf of Mexico: U.S. Environmental Protection Agency, accessed January 27, 2011, at URL http://www.epa.gov/bpspill/sediment-benchmarks html

U.S. Environmental Protection Agency, 2011c, Coastal water sampling, EPA response to BP spill in the Gulf of Mexico: U.S. Environmental Protection Agency, accessed January 27, 2011, at URL http://www.epa.gov/bpspill/water.html#understanding

U.S. Environmental Protection Agency, 2011d, National Recommended Water Quality Criteria, Appendix A—Conversion factors for dissolved metals: U.S. Environmental Protection Agency, accessed September 23, 2011, at URL http://water.epa.gov/scitech/swguidance/standards/criteria/current/index.cfm

U.S. Environmental Protection Agency, 2012, Approved general-purpose methods, accessed August 1, 2012, at URL http://water.epa.gov/scitech/methods/cwa/methods_index.cfm

U.S. Geological Survey, variously dated, National field manual for the collection of water-quality data: U.S. Geological Survey Techniques of Water-Resources Investigations, book 9, chaps. A1–A9, accessed September 23, 2011, at URL http://pubs.water.usgs.gov/twri9A

Wershaw, R.L., Fishman, M.J., Grabbe, R.R., and Lowe, L.E., eds., 1987, Methods for the determination of organic substances in water and fluvial sediments: U.S. Geological Survey Techniques of Water-Resources Investigations, book 5, chap. A3.

Wilde, F.D., and Skrobialowski, S.C., 2011, U.S. Geological Survey protocol for sample collection in response to the Deepwater Horizon Oil Spill, Gulf of Mexico, 2010, Sampling methods for water, sediment, benthic invertebrates, and microorganisms in coastal environments: U.S. Geological Survey Open-File Report 2010-1098, 178 p., accessed June 1, 2011, at URL http://pubs.usgs.gov/of/2011/1098/

Zaugg, S.D., Burkhardt, M.R., Burbank, T.L., Olson, M.C., Iverson, J.L., and Schroeder, M.P., 2006, Determination of semivolatile organic compounds and polycyclic aromatic hydrocarbons in solids by gas chromatography/mass spectrometry: U.S. Geological Survey Techniques and Methods, book 5, chap. B3, 44 p., accessed April 14, 2011, at URL http://pubs.usgs.gov/tm/2006/tm5b3/

Glossary

Censored data   A set of one or more censored values.

Censored value   An analytical result determined to be below a specified threshold concentration and reported as a less-than value (for example, less than 3 when 3 is the threshold; see Censoring).

Censoring   Application of a threshold to data such that concentrations above that threshold are quantified values and concentrations below that threshold are reported as less-than values (for example, less than 3, when 3 is the threshold). Censoring may be done at two stages of chemical data analysis: (1) by the laboratory because of high uncertainty in quantifying concentrations near the method detection limit, in which case the threshold is a reporting level, although for some methods, values that are below the reporting level but above the detection level could be quantified, but coded as estimates; (2) during data analysis because of blank contamination, in order to avoid interpreting incidental contamination in a sample as environmental contamination or to eliminate bias when comparing data, such as bias due to different method sensitivities, in which cases the threshold is a censoring level. This report describes three kinds of censoring levels: raised censoring levels, which are applied to contaminants detected in quality control blank samples; optimum censoring thresholds, which are applied for comparison of pre-landfall and post-landfall sample groups; and common detection thresholds, which are applied for comparison among contaminants with different laboratory reporting levels.

Censoring level   A concentration threshold that is applied to data such that concentrations above that threshold are quantified values and concentrations below that threshold are reported as less-than values (for example, less than 3).

Common detection threshold   A censoring level applied to a group of analytes for the purpose of comparing among the analytes. This eliminated bias due to differences in method sensitivities for different contaminants.

Detection level   A generic term for the lowest concentration that can be reliably quantified by a certain method at a certain laboratory.

Indeterminate sample   A sample with an indeterminate value for a specified analyte.

Indeterminate value   A censored value, that is, a result reported as less than a specified reporting level, where the reporting level is higher than the applied censoring threshold, so the value cannot be classified as either a detection or nondetection at that threshold. For example, an analyte concentration reported as less than 1 would be indeterminate at a censoring threshold of 0.2 because it is unknown whether the analyte is present at levels above 0.2.

Method detection limit   (MDL) The minimum concentration of a substance that can be measured and reported with 99-percent confidence that the value is greater than zero (40 CFR Part 136).

Optimal censoring threshold   The lowest censoring level that converts no more than 5 percent of results from censored to indeterminate values, maximizes the number of quantifiable detections and, if possible, also minimizes the number of indeterminate values. This was applied to concentrations of a given contaminant for the purpose of comparing pre-landfall and post-landfall sample groups.

Quantified value   An analytical result measured above the reporting level and reported as a specific concentration. For some methods, such as those in which there is corroborative evidence of analyte presence in a mass spectrogram, an analytical result measured below the reporting level but above the detection level would be quantified, but coded as an estimate.

Parameter code   Code for parameters in the USGS National Water Information System database (http://nwis.waterdata.usgs.gov/usa/nwis/pmcodes); also called USGS parameter code.

Raised censoring level   For contaminants detected in quality-control blanks, a censoring level higher than the reporting level that is applied to censor data in those environmental samples associated with the contaminated blanks, which minimizes the likelihood that detections of those contaminants in environmental samples are the result of incidental contamination. Typically, the raised censoring level is set at five times the maximum concentration determined in the applicable blanks or, for common laboratory contaminants, 10 times.

Reporting level   The concentration, set by a laboratory, used for reporting analytical results that are determined to be less than the detection level. This could be higher than the detection level because analytical results at or near the detection level can have high uncertainty. The reporting level can vary because of factors such as matrix interference, low sample mass, or sample dilution.

Appendix 1.   Methods, Reporting Levels, and Laboratories Used for Chemical Analysis for the Deepwater Horizon Oil Spill, Gulf of Mexico, 2010

Appendix tables and references are available for download in PDF format at http://pubs.usgs.gov/sir/2012/5228/.

Appendix 2.   Data Distributions for Contaminants in Water and Sediment Sampled in Response to the Deepwater Horizon Oil Spill, 2010

Appendix figures are available for download in PDF format at http://pubs.usgs.gov/sir/2012/5228/.

Appendix 3.   Benchmark Exceedances for Contaminants in Water and Sediment Sampled in Response to the Deepwater Horizon Oil Spill, 2010

Appendix tables are presented as Microsoft© Excel spreadsheets. They can be accessed and downloaded at http://pubs.usgs.gov/sir/2012/5228/.

Appendix 3.1.   Benchmark exceedances for organic contaminants in water, by sample, from the Deepwater Horizon oil spill, Gulf of Mexico, 2010

Appendix 3.2   Benchmark exceedances for organic contaminants in sediment, by sample, from the Deepwater Horizon oil spill, Gulf of Mexico, 2010

Appendix 3.3.   Benchmark exceedances for trace elements in water, by sample, from the Deepwater Horizon oil spill, Gulf of Mexico, 2010

Appendix 3.4.   Benchmark exceedances for trace elements in whole sediment and national baseline comparisons for trace and major elements and nutrients in the less than the 63-micrometer sediment fraction, by sample, from the Deepwater Horizon oil spill, Gulf of Mexico, 2010